GOVERNMENTALITY AND THE MASTERY OF TERRITORY IN NINETEENTH-CENTURY AMERICA

Matthew Hannah's book focuses on late nineteenth-century America, the period of transformation and upheaval which followed the Civil War and gave birth to the twentieth century. This was a time of industrialization and urbanization. Immigration was on the increase and traditional hierarchies were being challenged. In a sophisticated and fascinating study, Hannah explores the modernization of the American federal government during this period, using a rich tapestry of theoretical and empirical material. Discussions of gender, race and colonial knowledge are woven into an extended engagement with Foucault's ideas on 'governmentality' and other concepts from recent social theory. The empirical strands of the narrative are organized around the public career and writing of Francis A. Walker. A hugely influential figure at that time, he was director of the 1870 and 1880 US censuses, commissioner of Indian affairs and a prominent political economist and educator. Through an analysis of his work and his governing vision of social order, Hannah enriches and moves beyond previous interpretations of the period, demonstrating that the modernization of the American national state was a thoroughly spatial and explicitly geographical project.

MATTHEW G. HANNAH is Assistant Professor of Geography at the University of Vermont.

Cambridge Studies in Historical Geography 32

Series editors:
ALAN R. H. BAKER, RICHARD DENNIS, DERYCK HOLDSWORTH

Cambridge Studies in Historical Geography encourages exploration of the philoso-phies, methodologies and techniques of historical geography and publishes the results of new research within all branches of the subject. It endeavours to secure the marriage of traditional scholarship with innovative approaches to problems and to sources, aiming in this way to provide a focus for the discipline and to contribute towards its development. The series is an international forum for publication in historical geo-graphy which also promotes contact with workers in cognate disciplines.

For a full list of titles in the series, please see end of book.

GOVERNMENTALITY AND THE MASTERY OF TERRITORY IN NINETEENTH-CENTURY AMERICA

MATTHEW G. HANNAH

University of Vermont

CAMBRIDGE
UNIVERSITY PRESS

University Printing House, Cambridge CB2 8BS, United Kingdom

One Liberty Plaza, 20th Floor, New York, NY 10006, USA

477 Williamstown Road, Port Melbourne, VIC 3207, Australia

4843/24, 2nd Floor, Ansari Road, Daryaganj, Delhi - 110002, India

79 Anson Road, #06-04/06, Singapore 079906

Cambridge University Press is part of the University of Cambridge.

It furthers the University's mission by disseminating knowledge in the pursuit of education, learning and research at the highest international levels of excellence.

www.cambridge.org
Information on this title: www.cambridge.org/9780521669498

First published 2000

A catalogue record for this publication is available from the British Library

ISBN 978-0-521-66949-8 Paperback

For my parents, Claudia P. Hannah and John L. Hannah,
with love and admiration

Contents

List of illustrations		*page* x
Acknowledgments		xi
List of abbreviations		xiii
	Introduction	1
1	Governmentality in context	17
	Part I	41
2	The formation of governmental objects in late nineteenth-century American discourse	43
3	Francis A. Walker and the formation of American governmental subjectivity	60
4	American manhood and the strains of governmental subjectivity	84
	Part II	107
5	The spatial politics of governmental knowledge	113
6	An American exceptionalist political economy	160
7	Manhood, space and governmental regulation	188
	Conclusion	220
	Bibliography	229
	Index	240

Illustrations

1 Francis Amasa Walker *page* 150

2 Title page, *Statistical atlas of the United States*, 1874 151

3 Colored population, 1870 152

4 Foreign population, 1870 153

5 Native and colored population by states and territories, 1870 154

6 Constitutional population, 1870 155

7 Gainful occupations or attending school, 1870 156

8 Insane population, 1870 157

9 Principal constituent elements of the population of each state, 1870 158

10 Distribution of the population in accordance with temperature, 1880 159

Figure 1: J. Munroe, *A life of Francis Amasa Walker* (New York: Henry Holt and Co., 1923)

Figures 2 through 9: F. Walker (compiler) *Statistical atlas of the United States* (Washington, DC: Julius Bien, 1874)

Figure 10: Office of the Census, *Statistics of the Population of the United States at the Tenth Census*, F. Walker and C. Seaton (compilers) (Washington, DC, 1883)

Acknowledgments

Many people have provided me with invaluable help of one sort or another. If I have overlooked anyone in the following paragraphs, please accept my sincere apologies. The views expressed in this book, and the mistakes made, are entirely mine, and do not necessarily reflect (in fact sometimes directly contradict) the views or intentions of any of the individuals or institutions mentioned below.

For generous funding support, I would like to thank the University of Vermont's University Committee on Research and Scholarship, as well as the Dean's Fund of the College of Arts and Sciences. The decision on the part of Dean Joan Smith to institute a college-wide one-semester course reduction policy for junior faculty struggling to pursue research could not have been more timely. This research semester gave birth to Part I of the book, and allowed me to plan the rest.

For help with government documents and other primary sources, thanks go to Bill Gill, Government Documents Librarian, and Karen Campbell, Special Collections Librarian, both at the University of Vermont; Jeffrey M. Flannery, Manuscript Reference Librarian, Manuscript Division, Library of Congress; Bill Creech, Archivist, National Archives and Records Administration; Daria D'Arienzo, Archivist, Amherst College Archives and Special Collections; Becky S. Jordan, Special Collections, Iowa State University; the staff at MIT's Institute Archives and Special Collections and the staff at Johns Hopkins University's Special Collections Room. For tips and references to extremely useful secondary material, I would like to thank David Demeritt, Cynthia Enloe, Robert Nadeau, Jodi Pettazzoni, Subir Sinha, Robert Taylor, Rashmi Varma and Paul Wendt, as well as anonymous reviewers for Cambridge University Press.

For reading significant chunks (or all) of the manuscript or its predecessors at one time or another, and making excellent comments, I am grateful to Trevor Barnes, David Demeritt, Glen Elder, Deryck Holdsworth, Scott Kirsch, Anne K. Knowles, Joni Seager and Robert Taylor, as well as to anonymous reviewers for Cambridge University Press. Glen Elder deserves special thanks for dramatically altering my approach to the whole project in the course of a chance office conversation in 1996. David Demeritt and Robert Taylor made particularly crucial suggestions on a variety of points. Deryck Holdsworth, Vicky Cuthill and Marigold Acland have been extremely helpful in shepherding the project toward completion. Thanks to Sheila Kane for her discerning work at the copy-editing stage.

For hospitality at different stages of researching and writing, I am indebted to Amtrak, to Christa Ricken (without whom Chapter 3 would have been much more difficult), to Joni Seager and Cynthia Enloe (ditto Chapter 7), and to Brad, Irene and Rachel Wolf (revisions). Finally, to Antje Ricken, I owe deep gratitude for consistently benign neglect, and to Ulf Strohmeyer, for nothing and everything.

Abbreviations

DE	*Discussions in education*
DES	*Discussions in economics and statistics*
GPO	Government Printing Office
NARA	National Archives and Records Administration

Introduction

Between the end of the Civil War and the so-called "Progressive Era" that began roughly with the twentieth century, life in the United States was fundamentally transformed by massive industrialization, accelerating immigration, urbanization and the colonization of western North America. Many amid the welter of responses to modernization in this "Gilded Age" can be interpreted as expressions of what Robert Wiebe has called "the search for order."[1] The economic, political, social and cultural hierarchies inherited from the antebellum period, the constituents of "order," faced severe challenges. As a result of struggles by various groups and individuals to preserve, change or abolish the constituents of social order, some of its structures (e.g., the dominant economic position of countryside and commerce *vis-à-vis* city and manufacturing) were toppled and replaced, others (e.g., patriarchal gender relations) survived, albeit usually in altered form. One of the most important, if largely unintended, facets of industrialization and the quest for order in the modernizing West was an unprecedented expansion and diversification in the activities of national states.[2] Anthony Giddens and Michael Mann have done much to focus the attention of scholars on the importance, as well as the *spatiality* of this process.[3] Modern state formation has played a significant role in the shaping of modern societies, in part through the bounding and organization of national territories and populations into relatively integrated social, economic, political and cultural units.

There are two main purposes of this book: (1) to enrich our understanding of the historical geography of the modern American nation-state by approaching its history through Foucault's ideas about "governmentality"; and (2) to show that the workings of "governmentality" at a national scale are

---

[1] R. Wiebe, *The search for order, 1877–1920* (New York, 1967).
[2] M. Mann, *The sources of social power*, vol. II, *The rise of classes and nation-states, 1760–1914* (Cambridge, 1993).
[3] A. Giddens, *The nation-state and violence* (Berkeley, CA, 1987); M. Mann, *The sources of social power*, 2 vols.. (New York, 1986, 1993).

inherently and fundamentally spatial. Both goals will be pursued through an argument that brings certain key contributions to recent critical theory into sustained and detailed analytical contact with a specific geo-historical moment in the elaboration of modern power techniques. As I show how well an analysis centered on the concept of governmentality captures some of the complex features of an emerging logic of national power in late nineteenth-century America, I will simultaneously be demonstrating in great detail how the logic of governmentality is both fundamentally spatial and intricated with culturally specific concepts and ideologies. Thus the argument is neither essentially empirical nor essentially theoretical. I will neither simply test a theoretical framework against "the facts" nor merely shape an historical geography to conform to a pregiven theoretical structure. I am hoping instead to let each transform the other into something of more general interest and utility.

Like Giddens, Mann and other state theorists, Michel Foucault became keenly interested in the emergence of administrative techniques of national social ordering. Yet he brought to the topic his unique gift for careful analysis of the *logics* inherent in power techniques, and a healthy skepticism toward general stable categories like "the state," preferring on principle always to stress their *constructed* character. On the basis of lecture courses he gave at the College de France in the early 1980s, he and his collaborators embarked on a fine-grained, intensive exploration of key aspects of modern power. A collection of essays inspired by this course, published in English as *The Foucault Effect*, outlines an approach that illuminates many processes left unexamined in the more sweeping, expansive and implicitly essentialist accounts of Giddens and Mann.[4] In particular, Foucault's sensitivity to the mutual constitution of knowledge and power allowed him to give a nuanced account of epistemological aspects of state power. By weaving the problematic of governmentality into a geo-historically specific tale, it is possible to give real analytical bite to the superficially obvious claim that state formation must be *inherently geographical.* How did surveillance and regulation of national territories actually develop? Who were the advocates of governmentality, and who were its opponents? Which concrete social problems drove people to explore these new ways of governing? What did they think they were doing, and which cultural infrastructures, norms or belief systems inflected their programs? How did the quest for governmental control in fact transform American national space?

Even to ask such questions of Gilded Age America is a huge undertaking. To render the task more manageable, and to bring to life the various abstract logics I will be discussing, I shall weave the argument around the public career of Francis Amasa Walker (1840–97). A leading political economist, educator

---

[4] G. Burchell, C. Gordon and P. Miller (eds.), *The Foucault effect: Studies in Governmentality* (Chicago, 1991).

and social commentator, superintendent of the 1870 and 1880 censuses, commissioner of Indian affairs and many more things besides, Walker was probably the single most important early American proponent of what we would now call governmentality. Walker was a structurally and culturally "extensive" person: the web of subject positions he occupied and the range of ideologies and norms he brought to bear on American society were vast. Thus in tracing his work and thought, I will be able to say a good deal about the general logics, patterns, etc. animating his age. In addition, he was what we might call today a "closet geographer." His spatial habits of thought will prove immensely useful in what follows.

The components of Walker's worldview, however widely shared when considered one by one, came together in a more idiosyncratic way, and the more unique patterns of thought peculiar to Walker are also illuminating. In particular, it is possible to discern beneath the range of his opinions an unusually profound and abiding concern with "manhood." Whether he was discussing the workings of the American economy, the threat of immigration or the proper goals of education in an industrializing society, his rhetoric repeatedly betrayed an almost desperate urge to defend American manhood against the disruptions of the Gilded Age. Walker had strong opinions (as many of his contemporaries did) on other dimensions of social order, particularly race and immigration, but even there, a fear for besieged masculinity shaped his views. I will argue that his near obsession with manhood grew out of the most trying dimensions of his own experience: his involvement in the changing public affairs of the Gilded Age *stretched him beyond the limits of his almost legendary "manly vigor."* Walker's only biographer believed that the demands of public life, many of which concerned what I am calling governmentality, literally killed him.[5] Thus Walker's career will illustrate not only the spatial character of governmentality but also an important aspect of its cultural embeddedness. Gilded Age conceptions of American manhood were of course already undergoing massive change quite apart from the rise of the modern state.[6] But the rational mastery of national territory added a distinct set of problems to the mix. It should be clear by now that, in addition to *not* being a primarily empirical historical geography, and *not* being a primarily theoretical exercise, this book is *not* a biography. Although I will make no attempt to conceal my fascination with Walker in his own right, he functions for me as a *representative, a nexus of relations.* I am tempted even to suggest (perversely) that his individual uniqueness lies precisely in his "structurality," his "non-uniqueness." He was a singularly unquirky man.

There is, of course, an element of risk in using a single life to illustrate an

[5] J. Munroe, *A life of Francis Amasa Walker* (New York, 1923).
[6] M. Kimmel, *Manhood in America: A cultural history* (New York, 1996); E. A. Rotundo, *American manhood: Transformations in masculinity from the Revolution to the modern era* (New York, 1993).

important aspect of Foucault's thought; Foucault after all is the critic *par excellence* of the notion of a pregiven, unified individual subjectivity. But as I hope to show, Francis Amasa Walker, even with his constructed, polycentric, derivative status constantly in view, makes for an interesting tale.

Location in the intellectual landscape

In this study, I will attempt to demonstrate the advantages of an "intensive" rather than "extensive" use of theory. Rather than making sure I have cited every last mention of Walker in the historical literature, or every theorist who has ever made a contribution to our understanding of modern power, I will linger longer over a more limited range of documents and texts. In this also I follow Foucault, whose preference for selective but theoretically intensive, instead of comprehensive and theoretically cursory, historical analysis was a key part of what allowed him to produce such thought-provoking historical studies. I will dwell at length especially on the work of Foucault (obviously), but also draw occasionally on the work of Latour, James C. Scott and others.[7] Implicit in my use of their ideas is a belief that what Foucault termed "general" (as opposed to "total") history is still possible. As Foucault puts it, "A total description draws all phenomena around a single centre – a principle, a meaning, a spirit, a worldview, an overall shape; a general history, on the contrary, would deploy the space of a dispersion."[8] A set of general categories can allow comparisons that highlight differences and counterfactual alternatives as well as similarities and commonalities, that render intelligible struggles as well as conformity. This is, if you like, my "metatheoretical" commitment.

I am committed also to a particular "ethic" in exploiting Foucault's work. There are two general ways in which scholars positively disposed toward Foucault make use of him. On the one hand are arrayed the legions of critical researchers who *invoke* Foucault on a regular basis without necessarily accepting all facets of his approach to social explanation. This vast group of scholars most often puts him to work to signal their awareness of the importance of power relations in social life. Perhaps partly to counteract the cursoriness that can characterize this pattern of usage, a second and much smaller group of scholars, basing themselves in solid, careful readings of Foucault, has evolved an "exegetical" approach. In this style, one is careful to sort out exactly what Foucault did, and more crucially, did not say, and to clarify the

---

[7] M. Foucault, *The archaeology of knowledge* (New York, 1972); *Discipline and punish* (New York, 1977); "Governmentality," in Burchell, Gordon and Miller, *Foucault effect*, 87–104; Mann, *The sources of social power*, vol. II; B. Latour, *Science in action* (Cambridge, MA, 1987); *We have never been modern* (Cambridge, MA, 1993); J Scott, *Seeing like a state: How certain schemes to improve the human condition have failed* (New Haven, CT, 1998).

[8] Foucault, *Archaeology*, 10.

sorts of research to which Foucault's ideas are or are not applicable. This latter group is not composed of self-professed "Foucauldians"; in fact, many of them stress the perils of adhering to too many of his ideas. But they do effectively counterbalance the trivializing and reductionist tendencies of the first group.

What is missing from this picture can be imagined more clearly by comparison with the older and vaster literature on Marx. There we see not only the "invokers" and the "exegetes" but also a third group of scholars, who might be labeled the "elaborators". This group begins with a detailed reading of Marx, but moves from there to extend, to stretch, perhaps now and then to interpret the pregnant silences of Marx, all for the sake of uncovering Marx's more comprehensive relevance to historical and present day social analysis. While this group might be perceived by the exegetes to play a bit fast and loose with Marxism, it sustains a deeper engagement with Marx than do the fleeting invokers. I would like to see this study as one enactment of such a "third way" with respect to Foucault. It is an unashamedly Foucauldian study, but it is not limited by a "strict constructionist" reading of Foucault. This ethic should allow me to construct plausible articulations between Foucauldian and other types of analysis.

Since this book falls between the cracks of the usual genre distinctions, I should make a few further comments on its peculiarities. David Wishart's short but clear discussion of "the selectivity of historical representation" will serve as a useful touchstone.[9] Wishart distinguishes two stages in the process of historical representation, the identification of facts and the organization of factual statements into narratives. At each stage, the goal of "objectivity" is rendered permanently unattainable by the inevitable *selectivity*, first, of whatever evidence remains from "the events themselves," and second, of the cultural and individual "lenses" through which one interprets and orders this material. In my view, Wishart places too much stress on the idiosyncratic or subjective sources of the latter sort of selectivity, and too little on what might be called (unfashionably) "structural" patterns underlying the range of seemingly idiosyncratic uniquenesses. But be that as it may, the focus on sources of selectivity is valuable, and I shall say a bit more about those I can identify or anticipate in the present study.

First, by traditional standards of historical writing, this work will at first seem to give too strong a role to theoretical considerations. Even in the "empirical" section of the book (Part II), the narrative is organized not along a temporal sequence, nor around the salient events in Francis Walker's life and career, but rather around moments in an abstract logic of social control. Despite his awareness that *all* historical writing is implicitly suffused with

[9] D. Wishart, "The selectivity of historical representation," *Journal of Historical Geography* 23, 2 (1997), 111–118.

theory, Wishart takes a traditional tack in warning that "an imposed theory tends to change history [or historical geography] from a dialogue between the 'historian of the present and the facts of the past' into a monologue where the historian does all the talking."[10] I would counter with the observation that, like it or not, the historian *does* do all the talking. Every decision regarding how to understand, arrange and present historical material is a decision taken, in a more or less self-conscious, more or less "determined," way by the historian. Particularly in recent decades, it has become almost impossible for historical scholars to avoid the awareness that every aspect of their practice is suffused with theoretical assumptions. To be forthright about this, and thereby to have a more obvious presence in the narrative, is thus not at all to be narcissistic. Quite the opposite: by admitting to a particular, theoretically driven orchestration of selectivity, one can be more modest, and avoid the deceptive narcissism of "transparency." But all of this is not to say that one is no longer constrained by surviving evidence. It is simply to admit that while the evidence sets (always negotiable, and usually very wide) limits on the plausible, it is essentially *mute*, except at the most basic literal level of establishing that, for example, so-and-so many people died in a given year. The "larger" meaning of evidence is always supplied through the construction of narratives by interpreters, and there is always a range of possible "larger meanings." By itself, historical evidence is incapable of engaging in a "dialogue." As Mary Poovey might put it, Wishart's empowerment of evidence for dialogue is symptomatic of a credulous view of the "modern fact." The notion that "facts" can be both objectively available to observation and interpretable as evidence of unseen, abstract laws or tendencies has never been vindicated in modern thought. The connection between concrete particulars and systematic knowledge continues to require "something like a leap of faith."[11] The explicitness with which I link my various leaps to theoretical issues will, I hope, result in an argument that is neither needlessly timid nor overly assertive with respect to theory.

Having fully admitted my "bias" toward letting the theoretical questions shape the argument, it is nevertheless the case that Francis Walker's career and myriad pronouncements are uncannily well suited to illustrate the logic of governmentality. This logic does, for example, map rather well onto the passage of time and the succession of positions making up Walker's career. At times, the fit is truly astonishing, particularly in his ideas about social regulation (see Chapter 7). My intent in organizing the historical evidence has been to let the logic of social control, which is my primary interest, dictate the larger sweep of the presentation, and then to enfold this theoretically determined narrative in a cocoon or penumbra of running historical contextualizations. These contextualizations will usually involve reference to a few key secondary

[10] *Ibid.*, 115.
[11] M. Poovey, *A history of the modern fact: Problems of knowledge in the sciences of wealth and society* (Chicago, 1998), xix.

texts dealing with different aspects of late nineteenth-century American culture. In most cases these texts (many recent, some older) are widely acknowledged basic works in their respective fields, and have been influential in setting the terms of historical investigation since their publication. By linking my main narrative to these works, I intend deliberately to connect the problematic of governmentality to more established research problematics. However, I do not for the most part pursue the debates which have developed in these connected fields around the issues raised and claims made in the basic texts I cite. To do so would take me too far off the path dictated by a focus on governmentality, and would in any case be better done by scholars more qualified in these other fields. My use of secondary literature will thus appear eclectic, even thin, rather than comprehensive, in any one area. But this is perfectly consistent with the effect I would ideally like the book to have on its readers: an effect of provocation, a stimulation to recognize or at least consider associations, parallels or connections hitherto unnoticed or underappreciated. The point of what follows is not to settle anything in an historical sense, nor to establish as painstakingly as possible exactly what happened in some particular instance, but rather to provoke thought, including thought about the present. In this aim, I follow Foucault unabashedly. One of the most valuable but least appreciated aspects of his historical work was his privileging of the interesting over the safe. However modest the present effort, its allegiance is likewise to the interesting above all.

Substantively, this study significantly extends my work on the spatial aspects of modern techniques of social control. In the long run, I am after a relatively "general" (in the Foucauldian sense) theory of the spatial constitution of modern power relations: a sensitive set of geographical categories and concepts through which it will be possible to understand how a wide variety of concrete historical and present-day schemes of social control have operated (and been subverted through resistance). Foucault has remained near the center of my interests, both because of what he did write and because of what he left tantalizingly unexplored. My study of the US government's attempts to bring the Oglala Lakota (one of the two largest subgroups of the Sioux inhabiting the northern Great Plains) under control during the 1870s was intended to unearth the importance of the spatial conditions (particularly individual segregation and confinement) presumed but not systematically theorized by Foucault in his analysis of disciplinary power.[12] Forced to deal administratively with a semi-nomadic population that exhibited none of the spatial prerequisites of disciplinary power, the agents of the government did indeed make spatial fixation and separation major administrative goals (little remarked in the historical literature). Subsequently, I developed some of the insights from the historical study in a more theoretical discussion of the role

---

[12] M. Hannah, "Space and social control in the administration of the Oglala Lakota ("Sioux"), 1871–1879," *Journal of Historical Geography* 19, 4 (1993), 412–432.

of disciplinary power in constructing the day-to-day lives of "normal," spatially unconstrained modern citizens.[13] Most recently, I have brought the issue of *scale* into the picture, and have proposed a scale-based typology of different ways in which disciplinary social control operates in the modern world.[14] I begin with the familiar scale of single institutions and end with a discussion of "national discipline." But much of what I discuss under this last heading would have benefited from a more systematic theorization of the spatial underpinnings of governmentality. Governmentality is thus a "natural" next step in my attempt to piece together an understanding of fundamental relations between space and power. Few geographic studies yet exist with an exclusively governmental focus, but Jonathan Murdoch and Neil Ward have made an excellent start with their study of "the statistical manufacture of Britain's 'national farm'" during the early part of this century.[15] In particular, they show how central the collection of statistics is to the possibility of governmental control, and suggest that while governmentality is not simply large-scale discipline, the two techniques of power share important features. My attempt to present a more extended and systematic analysis of the spatial character of governmentality should be of interest not only to geographers but also to scholars in other fields who have found themselves in need of frameworks for understanding modern power relations.

Within the literature concerning the historical geography of state formation, this book relates most closely to the work of Felix Driver, especially to his *Power and pauperism*. In this and other pieces, Driver is clearly working to assemble a detailed picture of geographical aspects of state formation in Victorian Britain.[16] One of my goals is to bring the theory of governmentality more fully and systematically into this field of inquiry, and (by focusing on the US) to begin to make possible a more comparative, general understanding (in the sense of "general" discussed above). In *Power and pauperism*, Driver relies for most of his theoretical framework on Foucault's earlier analysis of discipline. He argues compellingly that the character and efficiency of a national system of disciplinary institutions will exhibit significantly uneven development due to the unevenness of the contexts in which it must operate. However, because the workhouse system was not clearly a *governmental* technique of

---

[13] M. Hannah, "Imperfect panopticism: Envisioning the construction of normal lives," in G. Benko and U. Strohmayer (eds.), *Space and social theory: Interpreting modernity and postmodernity* (New York, 1997), 344–359.

[14] M. Hannah, "Space and the structuring of disciplinary power: An interpretive review," *Geografiska Annaler*, 79 B, 3 (1997), 171–180.

[15] J. Murdoch and N. Ward, "Governmentality and territoriality: The statistical manufacture of Britain's 'national farm'," *Political Geography* 16, 4 (1997), 307–324.

[16] F. Driver, *Power and pauperism: The workhouse system, 1834–1880* (New York, 1993); "Moral geographies: Social science and the urban environment in mid-nineteenth century England," *Transactions of the Institute of British Geographers* 13 (1987), 275–287; "Political geography and state formation: disputed territory," *Progress in Human Geography* 15, 3 (1991), 268–280.

social control, Driver does not relate his findings in any extended way to the theory of governmentality. In the present work, I will focus on *unambiguously national* techniques of power, most saliently the national census and national immigration policy. Thus it will be easier to isolate a specifically national governmental logic.

Parts of my argument should also be of interest to geographers and other scholars who have recently begun to explore the power relations inherent in mapping and surveying. In my discussion of the census (Chapter 5), James C. Scott, Matthew Edney and others will be enlisted to help portray census taking, like the land surveys that usually precede it and the territorial mapping that often results from it, as an important geographical moment in the establishment of territorial mastery.[17] In principle, an adequate theorization of governmentality could in return provide a larger theoretical context for such research.

Finally, what of the larger, more traditional "isms"? The argument below is clearly informed by Marxism, feminism and critical studies of race and ethnicity. But since it is fundamentally a social constructionist argument, and thus attuned to complexity, the role played by each of the "big three" interpretive frameworks is limited by the other two, and by the complex effects of the interactions between class, gender and race, and other processes of contemporary concern. Taking "governmentality" as the basis for the analysis does not render the more familiar "isms" useless; in my view, they immeasurably enrich the meaning of governmentality. A Marxist analysis of class relations is the obvious background for relating governmentality to state formation (Chapter 1), and in helping to situate Walker's political economy with respect to late nineteenth-century American capitalism (Chapter 6). An awareness of gender issues makes it possible to recognize the importance of gender specificity both in Walker's self-understanding and in his prescriptions for the American social order (Chapters 4 through 7). Race and ethnicity comprise another key dimension of Walker's views on social ordering (Chapters 5 through 7). Taken together, Walker's views on these different basic issues gave him an understanding which can best be described as "American exceptionalist" (see end of Chapter 5). As an American exceptionalist, Walker wanted to understand the forces that threatened the stability of class, race and gender relations, and to use his understanding in the service of preserving America's uniqueness. Since he eventually linked problems in all three dimensions of social order to the same threat, it would make little sense for me to insist from the beginning that only one of the three dimensions was fundamental. On the other hand, I do make the argument that gender was for Walker the preeminent concern. I hope the narrative bears out this claim.

[17] J. C. Scott, *Seeing like a state: How certain schemes to improve the human condition have failed* (New Haven, CT, 1998); M. Edney, *Mapping an empire: The geographical construction of British India, 1765–1843* (Chicago, 1997).

Structure of the book and preview of the argument

The book is divided into two parts, the first setting the conceptual and historical frame, and the second focussing on Francis Walker's governmental program. In the first part, I explain what governmentality is and why it is a good way to approach the historical geography of state power techniques. Then I outline its concrete emergence as a discursive formation in Gilded Age America, and explain why Francis Walker is peculiarly well suited as a point of departure for its analysis. Here, Foucault's early archaeological writings prove extremely useful in assembling the range of subject positions Walker held together in his person. Next, I pause to relate governmentality (through Walker) to the contemporary culture of "manhood," both to illustrate the culturally embedded character of governmentality and to provide an important key to Walker's worldview. In the second half of the book, I take a long look at what Walker thought and did about the American social body. Having focused in Part II on how Walker was constructed by his time and place, I switch in Part II to a consideration of the *effects* that flowed from the actions of his admittedly derivative public persona. One point in doing this is to reinforce the notion that social constructionism does not disable causal analysis. In Part II, the fundamental spatiality of governmental power is demonstrated in a variety of ways, as is its intertwining with contemporary ideologies of race, class and gender. My central analytical argument is that a number of spatial issues lie at the heart of Walker's (and by implication, other) national programs of territorial mastery. These issues concern epistemological and regulatory politics of access to territory; different levels of tension between the need for fixity and the need for mobility among the people and resources occupying national territory; the centralization of control over social life, and the erection and contestation of boundaries at all scales from the household to the national territory as a whole. The book closes with an assessment of implications both for the study of governmentality and for the historical and present-day geographies of modern national power.

Chapter 1 explains Foucault's ideas on governmentality. This notion, as developed by Foucault, his collaborators and subsequent thinkers, is placed within a genealogical progression linking it with his earlier thoughts on "discipline" and "biopower." One of my purposes in proceeding this way is to call attention to the manner in which Foucault's enduring concern with knowledge and visibility was both sustained and altered as he moved from "disciplinary" to "governmental" matters. An important component of both disciplinary and governmental power is a general form of linkage between surveillance and regulation which can operate at a variety of spatial scales, and which I have termed elsewhere the "cycle of social control."[18] In formal terms, a cycle of

---

[18] Hannah, "Space and social control," 413.

social control links authoritative observation of an activity to (ideally) impartial judgment as to its acceptability, and (if necessary) to precisely calculated impartial penalization or correction of irregularities. Such a cycle can organize the minute regulation of individual activities in an institution, or control of larger groups of people and their activities over wide expanses of space (though of course the scale at which such control is exercised matters a great deal).[19] Its moments need not be concentrated in a single individual or even institution, as they are in this study. As I argue in Chapters 3 and 4, Walker lived in a time when national government was coming to be recognized as beyond the scope of any individual to comprehend. But analyzing it through, Walker renders its basic operation easier to see. Because this cycle is so central to the operation of governmental power, I will organize my discussion of Walker in the second part of the book around it.

The second major issue stressed in my presentation of governmentality is the notion of limits to knowledge and regulation. Whereas both disciplinary and governmental power share the "cyclical" structure I have just described, a major point of difference between the two is that the logic of disciplinary power recognizes no inherent limits to the scope for surveillance of individual lives. After spelling out this difference, I explore the points of contact and tension between a Foucauldian analysis of governmentality and certain forms of critical state theory. In contrast to the assumptions underlying many varieties of state theory, systems of governmental power involve governing authorities in a continuous process of trying to decide how much prying and regulating they should do. I situate this logic in part through a brief critique of the way recent scholars of the late nineteenth-century American state fail to capture it. Chapter 1 closes by noting the implied discursive, culturally embedded character and spatiality of governmental power.

Chapters 2 and 3 tackle the question of how the general principles of governmentality are first manifested in the emergence of an historically specific discursive formation. Foucault's template for understanding discursive formations is dusted off and put to work as a guide to the emergence of governmentality in Gilded Age America. After explaining Foucault's "archaeological" approach, I use part of it to focus attention on who it was that began to spin out a governmental discourse, which media and institutions they used as outlets, how intellectual authority was distributed in this early phase, etc. Chapter 2 begins with a discussion of the central role of statistics in governmental discourses, dwelling briefly on Theodore Porter's observations regarding the politics of statistical knowledge.[20] After explaining my decision to focus on the US Census, I offer a detailed look at the formation of governmental objects in late nineteenth-century discourse, organizing the discussion

[19] Hannah, "Space and the structuring of disciplinary power."
[20] T. Porter, *Trust in numbers: The pursuit of objectivity in science and public life* (Princeton, NJ, 1995).

around what Foucault calls "surfaces of emergence," "authorities of delimitation" and "grids of specification". In Chapter 3, I explore the details of the formation of governmental subjects (what Foucault terms "formation of enunciative modalities"), showing how the age of "best men" gave way to that of "experts." With a fairly comprehensive formal blueprint of the discursive formation in place, I then show just how well suited Francis Walker is to serve as a "governmental subject." He occupied the whole range of subject positions relevant to the governmental discursive formation, and was a key figure in anticipating certain emerging features of the incipient "Progressive Era." However, I do not treat the Progressive Era explicitly. This is partly because my aim (like that of other recent scholars) is to show how some practices commonly associated with it actually emerged earlier, and partly because Progressivism has already been so exhaustively documented. Even to outline the basic features of its literature would expand the scope of the present work significantly. Toward the end of Chapter 3, I begin to suggest that gender is a very important feature of his fitness for the role I give him.

Chapter 4 places this newly anointed governmental subject within a briefly sketched cultural history of American manhood. Industrialization, urbanization, immigration, etc. subjected the tough, independent, inventive American man who anchored contemporary gender ideology to serious stresses and strains.[21] The elite version of this ideal man, the omnicompetent public "man of affairs," faced a proliferation of ever larger, more complicated and intractable social problems, and found his powers of understanding and organization sorely taxed. Walker felt this change with particular acuteness, and in response attempted (as did some of his colleagues in public life) to mobilize the newly transformed "military" ideal of manhood that had emerged from the Civil War. His four-year stint in that conflict, which involved wounds, captivity and promotion through the ranks to (brevet) general, formed a central part of Walker's subsequent self-understanding. Addressed and referred to for the rest of his life by all who knew him as "General," Walker attempted to cope with the proliferation of responsibilities that hounded anyone aspiring to remain a "man of affairs" by means of a strict personal work regimen, constant employment of the considerable logistical skills he had honed in the army, and his reputation for honorable dealings. He was, in addition to all else, a respected amateur military historian, and much regarding his military manhood ideal can be gleaned from his writings on General Hancock (under whom he served) and on the Civil War more generally.

Many of the pressures exerted by modernization on the earlier robust, creative, resourceful, inventive and autonomous American ideal man can be summarized as a new imperative of self-discipline and training. If the post-Civil

---

[21] R. Slotkin, *The fatal environment: The myth of the frontier in the age of industrialization, 1800–1890*, (Middletown, CT, 1985); Kimmel, *Manhood in America.*

War ideal of military manhood represents a major first step toward incorporating this new imperative, the emerging figure of the scientist takes the logic further, issuing eventually (as we know) in a professional ethos of minute specialization and complete professional self-erasure. While relying primarily on the military ideal to understand his own life, Walker was at the same time a leading advocate of the formation of the modern social science disciplines that would help boost the scientist to a position of privilege. As first president of the American Economic Association, and through a long stint as president of the American Statistical Association, Walker worked actively to put an end to the possibility of careers like his own, and to usher in the "progressive" era of the expert. I argue not only that he was a key transitional figure, but also that the salience of the problem of manhood in his own experience led him to organize his entire understanding of American society around the issue.

Chapter 5 begins the second half of the book, which organizes the early workings of American governmentality around the three "moments" in the cycle of social control mentioned above: observation, normalizing judgment and regulation/punishment. To see how these three moments are linked is to see how governmentality actually works. Throughout the latter half of the book, it will become clear that early American governmentality operated in a very tenuous, incomplete and imperfect way. Nevertheless, it will also become clear that in so far as it works, governmentality must involve a complicated spatial logic which balances the conflicting demands for fixity (in the social body as an object) and mobility (of the agents through which this object is monitored).

In this chapter, I explore the process of observation at a national scale, focusing (as Walker and many of his contemporaries did) on the US Census as the foundation of early attempts at modern territorial administration. I place census taking alongside land surveys and mapping as a spatially organized epistemological tool with political implications.[22] The role of censuses in colonial administration is discussed in order to clarify these implications. The bulk of Chapter 5 is devoted to a detailed analysis of the spatial politics of census taking, which involves processes of abstraction, assortment and centralization as well as compilation. I will explain how they work with reference to Foucault's exegesis of the "panopticon," to Latour's discussion of "networks," to James Scott's treatment of modern state vision, and to Matthew Edney's perceptive analysis of the politics of cartography.[23] Francis Walker's struggles to improve the epistemological power of the 1870 and 1880 Censuses will illustrate many of the abstract claims made. I will return here to the issue of limits to governmental knowledge, in order to suggest that in practice, the

---

[22] J. B. Harley, "Deconstructing the map," *Cartographica* 26 (1989), 1–20; Edney, *Mapping an empire*.

[23] Foucault, *Discipline and punish*; Latour, *Science in action*; Scott, *Seeing like a state*; Edney, *Mapping an empire*.

significant limits are inherently spatial and often also gendered. This chapter closes with a look at Walker's 1874 statistical atlas of the United States, an internationally acclaimed, prize-winning effort, and the first national atlas based on census results in the Western world. Walker's understanding of his atlas as a representation of the American body politic expresses the spatial logic on the basis of which it was constituted.

Chapter 6 takes us from the "observational" moment of governmentality to the moment of "normalizing judgment" or implicitly "moral" assessment of observations. The inherent messiness and complexity of normalizing judgments at a national scale, the sheer variety of institutional sites from which they are offered and the democratic, federal structures through which they must be mediated have doomed this second moment in the cycle of social control to perpetual weakness and relative ineffectuality in most Western nation-states (especially the American).[24] The process of supposedly impartial assessment whereby knowledge gained from observation of the body politic is translated into socioeconomic policy remains highly imperfect today, despite the strenuous efforts of Walker and his spiritual inheritors through the twentieth century. The federal, relatively democratic structure of American government makes this inevitable: the logic of governmentality invests only some parts of a large, relatively decentered state structure, and is often neutralized or blocked by the workings of other parts and other logics. Nevertheless, there has been no shortage since the late nineteenth century of assessments of American social reality claiming to be objective and scientific, and we can learn much from them about how governmental knowledge has been organized in the interest of practical policy. Although few of Walker's judgments regarding whether and how to regulate society were carried through to the final policy phase of governmental power, they constitute an unusually complete and consistent overall social diagnosis.

Chapter 6 details Walker's anatomy of the American body politic, drawing on the wide range of books, magazine articles, speeches and official reports he published. The basic skeleton on which I will hang the discussion is his political economy. Here the issue was how to balance the mobility workers needed to compete successfully in the labor market with the fixity necessary for orderly social reproduction. As was typical of the times, Walker's political economy centered on the workings of capitalism but also embraced issues of demography, politics and morality. Some of his comments on contemporary social problems had nothing explicit to do with his political economy, but even so were almost always consistent with his systematic analyses. Walker's was of course but one among an infinite variety of systematic analyses of the

---

[24] M. Keller, *Affairs of state: Public life in late nineteenth century America* (Cambridge, MA, 1977).

principles of and threats to American social order, but almost every significant alternative opinion about economics, class, race, gender, immigration, urbanization or education found at least cursory treatment in his writings. Again, gender issues were a fundamental concern, but he also had an unusually geographical view of his society. Indeed there is ample basis for arguing (as I do) that the spatial order he longed for as superintendant of the Ninth and Tenth US Censuses played a role in structuring his view of the ideal American *social* order.

Chapter 7 explores Walker's thoughts (and actions) regarding regulation of the body politic (the final "moment" in the cycle of social control). It details, so to speak, the "actions" that issued from his observations and interpretations. It is here that Walker's thought is most distinctively governmental. In this chapter, even more than in Chapter 6, I take an active role in unearthing or reconstructing basic principles that can be used to tie together an overall picture of Walker's approach. A good portion of what I say here about his views would have been news to him, but again, the fundamental consistency I find in his opinions requires remarkably little interpretive "forcing" on my part. The two central themes in this chapter are (1) the limits to (and the nature of struggles against) social regulation (an important part of what differentiates governmentality from discipline), and (2) the selective advocacy of a range of spatial strategies for control. My substantive focus will be the two main pillars of his later social program: educational reform and immigration restriction.

There is also a consistent set of principles implicit in the range of spatial strategies Walker would like to see used to address different social problems. I will dwell at length on this aspect of his regulatory program, relating his prescriptions back to the analysis of the spatial logic of governmental knowledge given in Chapter 5. His previous experiences as commissioner of Indian Affairs will help spell out the range of options among which he had to choose in considering the problem of immigration. The tension between the various requirements Walker perceived for fixity and mobility will take on a special significance in helping to explain why the governmental principle of limited regulation often finds its strongest geographic expression (as it did with Walker) in programs of border control. The chapter ends with a detailed account of how Walker used his authority as a statistician to lay the groundwork for the comprehensive immigration restriction legislation that marked the early twentieth century. This episode represents a culmination both of Walker's career as a governmental subject and of my argument, as it illustrates so well the operation of all three moments of the cycle of social control in an actual system of governmentality.

The Conclusion returns to some of the basic questions raised here and in Chapter 1: How can theory improve an historical geographic analysis? What does the concept of governmentality offer to our understanding of the late nineteenth-century United States? How can a Foucauldian analysis of governmentality be improved in turn? What does it mean to say that governmentality involves at its roots a "spatial politics"? What can be made of the gendered dynamics identified in Walker's program? My comments will be brief and indicative.

1

Governmentality in context

In this chapter, I will spell out the particular construction of "governmentality" employed in the remainder of the study, and begin to place it in theoretical and empirical context. I explain the logic of governmentality at a fairly abstract and general level in the first section, with special emphasis on its relation to Foucault's earlier discussions of modern power techniques, and then address three issues that immediately arise. First, I begin to link the abstract treatment of governmentality to its concrete emergence as a discursive formation embedded in late nineteenth-century American life, and explain why Part 1 of the study will treat the discourses surrounding the United States censuses. Next I address the theoretical difficulties raised by the decision to pursue an analysis of governmentality within an empirical study centered on the American federal state. I outline a dialectical course by which it is possible to steer an interpretation between the seemingly nominalist implications of Foucault's thinking and the pitfalls of more "state-centered" studies, without violating the important insights of either. I then characterize the Gilded Age American federal state in a manner intended to highlight the points of compatibility and of tension between "micro" and "macro" approaches. Finally, I enumerate and briefly discuss the geographical issues inherent in any national governmental program, as a preview of the issues foregrounded in Part II.

Discipline and governmentality

Michel Foucault's thought has been of enduring interest to geographers particularly because his theorization of modern forms of power gives such a crucial role to space as a tool of social control. This preoccupation first emerged in his 1970s explication of disciplinary power. Foucault's understanding of disciplinary power has its classic expression in *Discipline and punish*.[1] Although the argument has been rehearsed both inside and outside

---

[1] M. Foucault, *Discipline and punish* (New York, 1977).

geography, its basic outline needs to be reviewed here.[2] This is because the principles by which disciplinary power operates remained central to Foucault's later explorations of governmentality. These later explorations were still at a relatively early and schematic stage when Foucault died in 1984, but the directions taken since that time by his colleagues and intellectual inheritors would be unintelligible in the absence of the groundwork laid by Foucault's earlier researches into modern power relations.

In his analyses of "discipline," Foucault documented the emergence, during the eighteenth and nineteenth centuries in Europe, of a new form of power different from those based on violence or law. This disciplinary power was rooted in visibility and surveillance, and involved the minute regulation and "normalization" of individual behavior through impartial observation and standardized, calculated punishment or correction of behavioral abnormalities. For Foucault, the ideal blueprint of disciplinary power was the "Panopticon" proposed by the English social reformer Jeremy Bentham as a model for English prisons. In the Panopticon, prisoners would be arranged in a multistoried ring of backlit cells around a central tower manned by watchers. The constant threat of visibility would, through an anonymous and impartial system of calculated punishments, encourage inmates to behave normally. At the same time, the detailed knowledge accumulated by prison authorities primarily for the purpose of control would be available also for use in the study of human physiology and behavior. As I have argued elsewhere, systems of control structured in this way can be seen to attach to individual activities what I call (adapting some of Foucault's concepts) "cycles of social control."[3] In an ideal panoptic system, each activity is (1) subject to constant threat of observation; and when observed (2) judged as to whether it is sufficiently "normal" or "regular," and finally, if not judged acceptable; it is (3) punished or corrected in an impartial, impersonal way. Although it takes very different forms in different situations and at different spatial scales, this cycle structures all forms of power organized around relations of vision. It will be helpful in the present study as a way to structure the historical analysis in Part II.

Foucault claimed that the basic panoptic logic, though it emerged unevenly and with considerable variation in different institutional contexts, and was

---

[2] F. Driver, *Power and pauperism: The workhouse system, 1834–1880* (New York, 1993), 10–15; "Bodies in space: Foucault's account of disciplinary power," in C. Jones and R. Porter (eds.), *Re-Assessing Foucault* (New York, 1992); G. Deleuze, *Foucault* (Minneapolis, MN, 1988); H. Dreyfus and P. Rabinow, *Michel Foucault: Beyond structuralism and hermeneutics*, 2nd ed. (Chicago, 1983); C. Philo, "Foucault's Geography," *Environment and Planning D: Society and Space* 10 (1992), 137–161.

[3] M. Hannah, "Space and social control in the administration of the Oglala Lakota ("Sioux"), 1871–1879," *Journal of Historical Geography* 19, 4 (1993), 412–432; "Imperfect panopticism: Envisioning the construction of normal lives," in G. Benko and U. Strohmayer (eds.), *Space and social theory: Interpreting modernity and postmodernity* (New York, 1997), 344–359.

always only imperfectly realized in practice, gradually and inconspicuously colonized many aspects of everyday life in the modern West. Notwithstanding the debates that have raged around every aspect of Foucault's argument, it would be difficult to imagine a clearer illustration of the mutual constitution of knowledge and power. Thus it is not surprising that Foucault's analysis of the Panopticon has acquired the status of an archetype in critical research on modern social control, or that "panoptic" has become a generic adjective.

Toward the end of the 1970s, Foucault began to extend and generalize his theorization of the mutual constitution of power and knowledge, and to explore the ways it operated at larger scales. His preliminary analysis of "bio-power" drew disciplinary power into a larger field, linking the regulation of individual bodies with larger-scale regulation of the "social body" through discourses of expertise regarding sexuality.[4] Biopower is here understood literally as "power over life." Foucault provides a concise overview of the historical emergence of biopower, and of its articulation with discipline:

In concrete terms, starting in the seventeenth century, this power over life evolved in two basic forms; these forms were not antithetical, however; they constituted rather two poles of development linked together by a whole intermediary cluster of relations. One of these poles – the first to be formed, it seems – centered on the body as a machine: its disciplining, the optimization of its capabilities, the extortion of its forces, the parallel increase of its usefulness and its docility, its integration into systems of efficient and economic controls, all this was ensured by the procedures of power that characterized the *disciplines*: an *anatomo-politics of the human body*. The second, formed somewhat later, focused on the species body, the body imbued with the mechanisms of life and serving as the basis of the biological processes: propagation, births and mortality, the level of health, life expectancy and longevity, with all the conditions that can cause these to vary. Their supervision was effected through an entire series of interventions and *regulatory controls: a biopolitics of the population*. The setting up, in the course of the classical age, of this great bipolar technology . . . characerized a power whose highest function was perhaps no longer to kill, but to invest life through and through.[5]

Sex and sexuality were particularly charged issues in this context because they were located "at the pivot of the two axes along which developed [this] entire political technology of life."[6] To attempt to regulate sexual activity was to link bodily disciplines with the management of populations. The empirical heart of the study is an exploration of the expert discourses that emerged to make these links. Foucault distinguishes "four great strategic unities which, beginning in the eighteenth century formed specific mechanisms of knowledge and power centered on sex": the "hysterization of women's bodies," the "pedagogization [subjection to careful guidance] of children's sex," the "socialization of procreative behavior" and the "psychiatrization of perverse pleasure." These

---

[4] M. Foucault, *The history of sexuality*, vol. I, *An introduction* (New York, 1978).
[5] *Ibid.*, 139. [6] *Ibid.*, 145.

four strategies targeted four objects of knowledge: the hysterical woman, the masturbating child, the Malthusian couple and the perverse adult. All four had to be managed in the interest of enhancing "life."[7]

Foucault intended his analysis as a critique of the Freudian "repressive hypothesis," the idea that, especially in the Victorian age, modern life has been accompanied by a fundamental suppression of sex and sexuality as topics of discussion and as necessary, legitimate and rewarding aspects of social existence. On the basis of his historical research, Foucault argues that, on the contrary, the nineteenth century *in particular* saw an explosion in public discussions of sex and sexuality.[8] He had originally planned to work backwards from the nineteenth century to investigate earlier regulatory and discursive practices surrounding sex (for example, the practice of confession). But he decided to take a different tack, and the last two published volumes of his history of sexuality concerned practices of sexual self-constitution in Greek and Roman antiquity.[9] In his explanation of the switch, Foucault made clear that he had not changed course because there was anything fundamentally "wrong" with the focus on biopower; it was only that it led him into a line of questioning which diverged from his larger purpose in writing a history of sexuality.[10] However, it was not until much later that many in the Anglophone world realized that he had also continued to pursue his interest in "methods of power capable of optimizing forces, aptitudes, and life in general without at the same time making them more difficult to govern," and that this had led him to conceptualize "governmentality."[11]

Before moving on to governmentality as such, it is worth dwelling at greater length on the earlier theorization of biopower. Many of the issues I will treat in the process of contextualizing governmentality in the late nineteenth-century United States are already foreshadowed here. The first is that of gender. Although Foucault does not develop the point very far, it is fairly obvious that the fourfold regulation of sexuality traced in the earlier study meshes easily with a patriarchal, paternalistic gender ideology. While the four "targets" of regulation included men as well as women, the experts doing the regulating were (as they still are) in practice mostly men, and the purpose of the four strategies was to bolster the patriarchal nuclear family as the lynchpin of social order and health. Given this set of circumstances, a general ideological "masculinization" of key aspects of social regulation is not difficult to imagine. Foucault's analysis of biopower renders more intelligible, for example, the sort of link between regulation and manhood drawn so vividly by Francis Walker (see Chapters 4 and 6).

Second, Foucault recognizes in this early study that a logic like that of bio-

---

[7] *Ibid.*, 103–105.   [8] *Ibid.*, 8–13.

[9] M. Foucault, *The history of sexuality*, vol. II, *The use of pleasure* (New York, 1986); *The history of sexuality*, vol. III, *The care of the self* (New York, 1988).

[10] Foucault, *The use of pleasure*, 4–6.   [11] Foucault, *The History of Sexuality*, vol. I, 141.

power is never nested in a cultural context as a "pure," self-contained island of discourse and practice. Rather, it is always shot through with residues and traces of other external ideologies, perhaps survivals from earlier moments in the history of particular cultures. This point comes out most clearly in Foucault's discussion of the survival (throughout the nineteenth century) of an older "thematics of blood" connected with the death-oriented ideologies of social order (violence, purity, inheritance, breeding) that had preceded the new concern with life:

Beginning in the second half of the nineteenth century, the thematics of blood was sometimes called on to lend its entire historical weight toward revitalizing the type of political power that was exercised through the devices of sexuality. Racism took shape at this point (racism in its modern, "biologizing," statist form): it was then that a whole politics of settlement (*peuplement*), family, marriage, education, social hierarchization, and property, accompanied by a long series of permanent interventions at the level of the body, conduct, health, and everyday life, received their color and their justification from the mythical concern with protecting the purity of the blood and ensuring the triumph of the race.[12]

This use of biopolitical means for racial ends is perfectly illustrated by Francis Walker's involvement in the American movement for immigration restriction (see Chapters 6 and 7).

The final foreshadowing worth brief mention here concerns the relation between social regulation and the requirements of industrial capitalism. Foucault remained wary until the end of "grand narratives" such as that of Marxist history, and for very solid reasons. But in his 1978 work we can glimpse the possibility of an articulation between his supple "non-totalizing" analyses and more sweeping narratives, or at least an acknowledgement that the two approaches need not always preclude each other:

[B]io-power was without question an indispensable element in the development of capitalism; the latter would not have been possible without the controlled insertion of bodies into the machinery of production and the adjustment of the phenomena of population to economic processes. But this was not all it required; [capitalism] also needed the growth of both these factors, their reinforcement as well as their availability and docility . . .[13]

Although he would never go out of his way to stress the possibility of common ground, his work on governmentality would multiply the opportunities to connect his perspective with one centered more on general categories such as "mode of production" or "the state" (see below, p. 32). Some sort of connection would in any case have been impossible to avoid, given that the analysis of governmentality has at its core an analysis of discourses of political economy.

[12] *Ibid.*, 149.   [13] *Ibid.*, 141.

This extended stopover in Foucault's discussion of biopower should bring home the point that any such technologies of control are unavoidably intertwined with many different sorts of ideological and material mediations. These mediations originate in the "external" cultural and political environment into which new logics of social control emerge, but they may nevertheless play fundamental roles in determining precisely *how* social regulation operates in concrete, geo-historical practice.

The 1991 publication of *The Foucault Effect* made many readers in the anglophone world aware for the first time that Foucault had not simply dropped the study of the history of social regulation around 1980 to pursue the new line of inquiry into "techniques of the self" that led to his last two book-length studies.[14] His essay "On governmentality" was published in English as far back as 1979, but it must have languished relatively unremarked through most of the 1980s, since there were few anglophone responses to it before 1990.[15] As the 1991 collection reveals, in tandem with his work on practices of sexual self-constitution in antiquity, Foucault and a number of European collaborators continued to flesh out the history of biopower. The result was a more nuanced general history of larger scale social regulation, and, of particular importance to this study, a key insight into the way the rise of "liberal" political economy qualified the Western practice of biopower.

Governmentality has a range of meanings, denoting both a general analytical category and more historically specific forms of power that manifest some or all of the features of an abstract logic.[16] In the specific historical sense in which I will employ it, it means roughly what I have been calling large-scale biopower, but with a more explicit connection drawn between demographic and economic trends and processes. Governmentality is "the ensemble formed by the institutions, procedures, analyses and reflections, the calculations and tactics that allow the exercise of this very specific albeit complex form of power, which has as its target population, as its principal form of knowledge political economy, and as its essential technical means apparatuses of security [modern institutions for the improvement and administration of life]."[17] In an excellent and influential essay, Nikolas Rose and Peter Miller provide an overview of the full range of phenomena that can be understood to fall under the category of governmentality.[18] These phenomena range in scale from individual "self-help" initiatives to life insurance provison to the policies of nation-states. My usage of the term will be much more restricted, owing in

---

[14] G. Burchell, C. Gordon and P. Miller (eds.), *The Foucault effect: Studies in governmentality* (Chicago, 1991).

[15] M. Foucault, "On governmentality," *Ideology and Consciousness* 6 (1979), 5–21.

[16] M. Foucault, "Governmentality," in Burchell *et al.*, *The Foucault effect*, 102 103; C. Gordon, "Governmental rationality: An introduction," in *ibid.*, 3.

[17] Foucault, "Governmentality," 102.

[18] N. Rose and P. Miller, "Political power beyond the state: Problematics of government," *British Journal of Sociology* 43, 2 (1992), 173–205.

part to the nature of my empirical concerns and in part to my commitment to explore the scope for common ground between Foucault and other theorists. The principal restriction will be an almost exclusive focus on programs and institutions explicitly geared toward national-scale social regulation. I take as one cue for my narrower usage the additional claim Foucault makes in the same essay that governmentality has tended "to predominate over sovereignty and discipline, and to generate a whole complex of knowledges" in the modern era.[19] This clearly implies that although governmentality also involves surveillance, it is something other than merely large-scale discipline or the aggregate system of disciplinary institutions. For Foucault, "we need to see things not in terms of a replacement of a society of sovereignty by a disciplinary society and the subsequent replacement of a disciplinary society by a society of government; in reality one has a triangle, sovereignty – discipline – government, which has as its primary target the population and as its essential mechanism the apparatuses of security."[20] Governmentality and discipline are articulated but not identical. Governmentality, like discipline, constructs (not merely "manipulates") its objects, but unlike discipline, it constructs them as objects that should not be unduly manipulated. I will be concerned with it chiefly as a national form of biopower that more or less successfully infuses state institutions and their behavior. This may seem a problematic way to identify governmentality, since I treat the "state" as an established fact in a manner generally avoided by Foucault. Yet if we return to the genealogy of governmentality traced by Foucault and his collaborators, it will be apparent that the national state is presupposed as an important context for the basic logic of power. Since this issue has such important implications for the possibility of linking Foucauldian to other sorts of analysis, I will return to it again at greater length below.

With the concept of governmentality Foucault and his collaborators in effect replaced the earlier, monolithic notion of a "biopolitics of population," a discourse and practice of social regulation which emerges in the eighteenth century and persists (with minor variations) until the present, with a more finely differentiated series of stages in social thought. In the genealogy these theorists have constructed, the key insight which marks specifically governmental thinking can be credited to the "physiocratic" theorists of eighteenth-century France, who realized that the regularities displayed by social statistics implied the existence of a realm of social reality outside the state, a realm possessing its own independent laws of development and behavior.[21] But in keeping with inherited assumptions about state knowledge as the self-knowledge of the sovereign, the physiocrats continued to believe that this socioeconomic realm could be completely known by the state. According to Foucault,

---

[19] Foucault, "Governmentality," 103.   [20] *Ibid.*, 102.

[21] Foucault, "Governmentality," 99; G. Burchell, "Peculiar interests: Civil society and governing 'the system of natural liberty'," in Burchell, *et al.*, *Foucault effect*, 126.

Adam Smith's notion of the "invisible hand" represented the crucial next step in the development of Western thinking about social regulation. It codified the foundation of the liberal view, in which, for the first time, social and economic processes were seen to be opaque to the sovereign, and were seen to operate according to an independent set of laws. For these reasons, the state had to be dissuaded from attempting comprehensive regulation of society.[22] "The advent of liberalism coincides with the discovery that political government could be its own undoing, that by governing over-much, rulers thwarted the very ends of government."[23] In short, the relationship between state and society had come to be seen, by the late eighteenth century, as one of exteriority, and the character of "society" as quite complicated, involving people in their relations with "wealth, resources, means of subsistence, the territory with its specific qualities, climate, irrigation, fertility, etc. . . . [with] customs, habits, ways of acting and thinking, etc . . . [and with] accidents and misfortunes such as famines, epidemics, death, etc."[24]

According to the wisdom of liberal political economy, rulers could only prosper if this complex web of people, things and processes prospered, but this would be possible only if the state allowed, encouraged or facilitated the unhindered operation of socioeconomic laws that governed "civil society," laws which it could not completely understand or control. There was still a definite role for knowledge as an instrument of power, but no longer through direct and comprehensive manipulation. On the liberal view, knowledge had to provide a basis not only for programs of state action, but prior to that, for decisions about whether state action would be appropriate at all.[25] In order to deal with these demands, the state had not only to preserve society and economy, but also to "ensure the existence of political spaces within which critical reflections on the actions of the state are possible . . . [i.e.] observe and maintain the autonomy of the professions and the freedom of the public sphere from political interference."[26] In the logic of liberal thought, the social sciences came to play a pivotal role, because they "provide[d] a way of representing the autonomous dynamics of society and assessing whether they should or should not be an object of regulation."[27]

To sum up, then, governmentality, like discipline and like other forms of biopower, is at this general level a rationality of social control based in the mutual constitution of knowledge and power. It constructs an object of knowledge, the social body, through discursive practices which, in giving it

---

[22] Burchell, *Peculiar interests*, 134; Gordon, "Governmental rationality," 20; T. Porter, *The rise of statistical thinking, 1820–1900* (Princeton, NJ, 1986), 17.

[23] A. Barry, T. Osborne and N. Rose, "Introduction," in A. Barry, T. Osborne and N. Rose (eds.), *Foucault and political reason: Liberalism, neo-liberalism and rationalities of government* (Chicago, 1996), 8.   [24] Foucault, "Governmentality," 93.

[25] T. Osborne, "Security and vitality: Drains, liberalism and power in the nineteenth century," in Barry, *et al.*, *Foucault and political reason*, 101.

[26] Barry, Osborne and Rose, "Introduction," 10.   [27] *Ibid.*, 9.

intelligible form, render this object at least partially susceptible to rational management. As such, governmentality, too, is fundamentally structured around cycles of social control linking observation, normalizing judgment and regulation. But unlike its genealogical predecessors, governmentality involves a more complicated version of the cycle. Observation has two purposes: it not only provides a means of comparing some construction of social reality with a norm; it also helps the governing authorities decide whether there are limits to its ability to enforce or achieve the norm, whether the attempt should be made to correct any perceived deviations. This more circumspect approach to regulation goes hand in hand with an attitude toward the social body of care, cultivation and enhancement, an attitude less evident in the exercise of disciplinary power. While the disciplines are also designed to enable, their point is less to enable any pre-existing interests or goals of the individuals they "subject," but rather to enable non-disruptive integration into an externally defined social order, whether or not such integration is desired by those subject to it. In short, governmentality at a national scale involves more respect for the integrity and autonomous dynamics of the social body.

The questions raised by the notion of governmentality can be classed in three broad categories. They all have to do in different ways with context. (1) How does governmentality actually emerge in geo-historically specific social settings (how does its advent affect, and how is it affected by, the specifics of these settings)? (2) To what extent is it theoretically permissible to weave a Foucauldian view of governmentality into an empirical study centered on the activities of "the American federal state"? (3) What are the *geographical* issues inherent in national-scale programs of governmentality? These questions are my overarching concern in the remainder of this book. Here I will lay out their basic coordinates within the context of late nineteenth-century America.

The concrete emergence of American governmentality as a discourse

What sort of phenomenon is governmentality, in concrete terms? As a "rationality" or a "logic," it is first of all a discourse (or set of discourses). Recall that the discourse of political economy is at the heart of Foucault's definition. Its broad purpose is to persuade those who govern to do so according to the principles of political economy (for example, to move cautiously in regulating economies). But this is only a general guideline. How does the state actually learn about (i.e., construct) society in order to decide just how to govern it? What Michael Mann notes about bureaucracies in the West was probably true in general of every aspect of the expansion of modern power techniques during the late nineteenth century: they were "everywhere preceded by [their] ideologies."[28] Most governmental measures had been advocated for some time

---

[28] M. Mann, *The Sources of social power*, vol. II, *The rise of classes and nation-states, 1760–1914* (New York, 1993), 472–473.

before they were put into wide-spread practice. Indeed, as Rose and Miller argue, "[w]e do not live in a governed world so much as a world traversed by the 'will to govern,' fuelled by the constant registration of 'failure,' the discrepancy between ambition and outcome, and the constant injunction to do better next time."[29] This was all the more true in the late nineteenth-century United States, where most proposals for the implementation of national-scale governmental programs were never operationalized.

In a crucial contribution to *The Foucault effect*, Giovanna Procacci argues that governmentality has actually involved two distinct levels of discourse: the familiar but relatively abstract and deductive level of political economy, and the more mundane practical level of "social economy."[30] She begins with the problem of "pauperism" (chronic poverty), which posed an unsolvable puzzle to political economists in the early nineteenth century. Pauperism confronted them with the spectacle of people whose condition should have been impossible given the contemporary wisdom that free markets and the invisible hand were universally beneficial and redemptive social forces. Paupers were living, breathing proof that leaving economies alone was an insufficiently subtle way of accommodating independent economic dynamics. In parallel with political economy, there accordingly arose another level of discourse, inspired by the more empirical German tradition of historical economics. "Social economists" insisted on confronting and studying the *reality* of pauperism. Their discourse is portrayed by Procacci as a "*savoir* . . . mediating between the analytico-programmatic levels of the sciences [political economy and its deductive approach] and the exigencies of direct social intervention."[31] This *savoir* (as opposed to the more self-consciously scientistic "*connaisance*" of political economy) "relocates the object thus scientifically delineated within a field of relationships in which the instruments of the scientific project [in this case, mostly deductively derived categories and the abstract relationships between them] are forced into contact with all the rigidity, inertia and opacity which the real displays in its concrete functioning."[32] This definition characterizes a general approach, which applies to many other issues in addition to pauperism, and which will be very helpful in making sense of Francis Walker's efforts. Two features of social economy are particularly relevant to the story I tell below: (1) unlike political economy, social economy is necessarily and openly

---

[29] Rose and Miller, "Political power beyond the state," 191.
[30] G. Procacci, "Social economy and the government of poverty," in Burchell *et al.*, *Foucault effect*, 151–168.   [31] *Ibid.*, 156.
[32] *Ibid.*, 157. Rose and Miller distinguish *three* levels of governmentality: "rationality," "program" and "technology," the last of these embracing the insciptions, recordings and constructions of social problems that go on in various locales of expertise located outside state apparatuses but articulated with them. Since the technologies I will be investigating are *state* technologies more tightly associated with "programs," I will treat Procacci's "social economy" as a category including both "programs" and "technologies," and "political economy" as roughly equivalent to what Rose and Miller term "rationality."

concerned with morality because its object is concrete socioeconomic (not merely economic) order; (2) social economy gives a prominent role to statistics, which in the political economy of the period tend to appear only as supports for deductive conclusions. Both of these features imply that social economy cannot simply rely on *laissez-faire* principles as guidelines for deciding what not to regulate. Governmentality, as *savoir* and *connaisance,* is something more subtle than blind reliance on this one (in)famous liberal axiom. It is in making such subtlety possible that censuses and other statistical forms of knowledge play their most important role.

Procacci's distinction dovetails (albeit imperfectly) with certain aspects of Mary Poovey's analysis of "the modern fact."[33] The modern fact has, according to Poovey, been understood as simultaneously an objectively observable concrete particular and a piece of "evidence" for unseen laws (whether natural or social). But Western thought has never succeeded in certifying that knowledge of facts can be connected successfully in accurate accounts of laws. After the failure of a long series of more or less explicit philosophical attempts to solve this problem, Poovey argues that it was finally tamed, though not solved, by the formalized separation of fact-gathering and systematic interpretation into different professions ("statistics" and "political economy").[34] To put it a bit more bluntly than Poovey would, if political economists could disown responsibility for the accuracy of facts, statisticians could disown responsibility for the way they were interpreted. The persistence of the fundamental epistemological problem with the modern fact could be blurred by "sleight of discipline." Liberal governmentality, consisting as it did of an articulation of political economy and statistics, incorporated the problem of the modern fact without solving it. The connection between facts and systematic knowledge is constructed, according to both Poovey and Procacci, through morality.

This means that its discourses necessarily become entangled with ideological components of the larger culture which relate to the definition of proper social order. In his discussion of biopower and sexuality, Foucault foreshadowed two specific entanglements of this kind: with gender and racial ideologies. Along with the notion of class already central to political economy, these two issues structured the most common generic understandings of social order in Victorian American culture. Francis Walker, comprehensive "governmental subject" that he was, not only involved himself prominently in debates at both levels of discourse, but also in all three wider ideological issues. I will treat his views on class and race at some length in Chapters 6 and 7, but it is my contention that gender was for him the single most important dimension of social order in one crucial sense. While race was also a key dimension in which Walker came to understand social order in an abstract sense, he

---

[33] M. Poovey, *A history of the modern fact: Problems of knowledge in the sciences of wealth and society* (Chicago 1998).   [34] *Ibid.*, 304–305.

connected problems of social order to concrete individual lives and activities through the discourse of manhood. Like most social commentators of his age, Walker was steeped in individualism, and thus could not conceive of larger social forces or phenomena entirely apart from the individual actions through which they were expressed. Yet for Walker, the individual was most primordially a *gendered* being. Thus Chapter 4 will be devoted to understanding how he became so concerned with "manhood," and how this concern colored his interventions in the discursive formation I will have explained in Chapters 2 and 3. My claim is not that governmentality *must* be structured fundamentally by gender issues, but rather (1) that we should not be surprised to find that it often is, and (2) that through an instance in which it was, we can learn something more general about how governmental rationality interacts with substantive ideologies of social order.[35]

At the most basic level, the possibility of national governmental objects presupposes the possibility of national phenomena, and in this sense is predicated on the (at least imagined) existence of a national state. Objects must also be constituted so as to be (in principle, at least) *rationally manipulable,* that is, manipulable in a way that avoids the appearance of arbitrariness. Crucially, this requires that it be possible to define with some precision what state the "social body" is in at any given time, that it be possible also to define more or less precisely some standard of "good condition" against which its actual condition can be compared, and that it be possible to determine on the basis of non-arbitrary grounds whether intervention in the workings of society is justified in any given case. These general rules of formation strongly favor the constitution of governmental objects on the basis of some kind *of quantitative measurement scheme.* This is fairly obvious in the case of rational manipulability: "[q]uantification is a way of making decisions without seeming to decide," a way of exercising "power minus discretion."[36] But numbers are also "particularly well suited for communication that goes beyond the boundaries of locality and community."[37] Bruno Latour puts the problem of rational, national manipulation nicely: "how to act at a distance on unfamiliar events, places and people? Answer: by somehow bringing home these events places and people. How can this be achieved, since they are distant? By inventing means that (1) render them *mobile* so that they can be brought back; (2) keep them *stable* so that they can be moved back and forth without additional distortion, corruption or decay, and (3) are *combinable* so that whatever stuff they are made of, they can be cumulated, aggregated, or shuffled like a pack of cards."[38] Numbers fit the bill nicely.

Some caution should be exercised here to avoid naturalizing statistics as the

---

[35] See Jacques Donzelot, *The policing of families* (Baltimore, MD, 1997 [1979]).

[36] T. Porter, *Trust in numbers: The pursuit of objectivity in science and public life* (Princeton, NJ, 1995), 8, 98.   [37] *Ibid.*, ix.

[38] B. Latour, *Science in action* (Cambridge, MA, 1987), 223.

transhistorical *telos* of representations of social reality. In her *History of the modern fact*, Mary Poovey makes it clear that the process by which numbers came to be associated with "objectivity" in social science was torturous, indirect, contingent and always reversible.[39] Among other things, the political economists who finally succeeded in placing statistics at the core of modern social science in the early nineteenth century had to stabilize a dubious association between precision and accuracy, forge a link between numbers and the credibility of their users, overcome the widespread suspicion of the "amorality" of numbers that had led to political economy's stigmatization as "the dismal science," and invest an attitude of "disinterest" with positive epistemological value. Poovey suggests that this rickety, provisional conceptual network comprising the modern fact has long been in decline, eclipsed by the "postmodern" fact based in virtual modeling. In my view, this judgment is a bit hasty. The ease with which we can accept Latour's and Porter's explanations for the usefulness of statistical representation indicates that the modern fact is still alive and well. Poovey's point that the primacy of statistics is an historical, contingent phenomenon is well taken, but the present study is still located within the modern phase of epistemic history, and therefore makes use of some of its assumptions.

A truly comprehensive account of national-scale, late nineteenth-century American governmentality would have to include in its purview all significant metrical schemes that met these criteria, for example the early eastern and later, more spectacular western land surveys; the township and range system and other grids of land division; the various mineral, agricultural and forest surveys; the anthropological inventories begun by John Wesley Powell; and all the cartographic projects undertaken to order the information provided by these systems of observation.[40] I will confine myself in this study to the nineteenth-century system of "social statistics," but many of the issues introduced here will be more generally relevant to other national measuring schemes (see Chapter 5).

An American fascination (particularly in the nineteenth century) with social statistics has been noted by a number of scholars, and is given an interesting explanation by Theodore Porter.[41] Porter begins within the modern era of the fact identified by Poovey, viewing statistics as a "strategy of communication," and argues that statistics will tend to predominate over other forms of communication wherever there is a need for specialized knowledge coupled with a strong public distrust of secretive expertise, and wherever specialists are

---

[39] Poovey, *History of the modern fact*.

[40] A. H. Dupree, "The measuring behavior of Americans," in G. Daniels (ed.), *Nineteenth century American science: A reappraisal* (Evanston, IL, 1972), 22–37.

[41] P. C. Cohen, *A calculating people: The spread of numeracy in early America* (Chicago, 1982); R. C. Davis, "The beginnings of American social research," in Daniels (ed.) *Nineteenth century American science*, 152–178; Porter, *Trust in numbers*.

vulnerable to the consequences of distrust. This pattern is especially apparent in the United States, which nourishes a long tradition of distrust of expertise (part of a more general tradition of anti-intellectualism), and where bureaucracies, scientific disciplines and other authoritative organizations are not well insulated from public displeasure.[42] Especially in the early nineteenth century, descriptive statistical knowledge was thought comprehensible enough to be accessible to many Americans, and thus to be an excellent basis for decision making in the context of a relatively egalitarian democracy. Yet the need for specialization was still relatively modest at that time; as the century wore on, the transparency of authoritative organizations *and their numbers* would decrease.

There were three main streams of statistically based social research in the period before the Civil War: the first and most important component of the "statistical movement" developed around the "administrative needs of the state and the economy and found expression mainly in the Census and other official statistics"; the second stream centered on vital statistics stemming from the epidemiological research of medical doctors and the demographic calculations of insurance actuaries; the third stream concerned "moral statistics" gathered in the effort to tackle urban social problems.[43] The latter two streams of the statistical movement were not at first national in scope. The discourses of insurance, spurred in part by the occasional outbreak of epidemics and in part by ongoing international debates about the salubrity of the American climate and the vigor of its animal and human life, nevertheless lacked even a rudimentary nationwide life table until 1868. Early "moral statistics" on crime and prisons, pauperism, insanity and other social pathologies, were even more local in character, rarely extending up to the level of individual states.[44]

To see how a national entity suited to governmental management was statistically constituted in the years following the Civil War we must focus on the first of Davis's three streams: national governmental statistics. From the earliest years of the republic, there were two main kinds of regularly collected national statistics: statistics of international trade, and decennial census statistics.[45] In a sense, they complemented each other: international trade numbers helped define the nation externally as an aggregate economic entity distinct from others with which it interacted, while census statistics defined it internally. Since the latter register the formation of social as well as narrowly economic objects, I will focus in the remainder of this study on the nineteenth-century US Census. During the Gilded Age, the US Census "became a full-

---

[42] R. Hofstadter, *Anti-intellectualism in American life* (New York, 1962); Porter, *Trust in numbers*, 6–8.   [43] Davis, "The beginnings of American social research," 153.

[44] *Ibid.*, 167–170, 171–175.

[45] J. Cummings, "Statistical work of the Federal Government of the United States," in J. Koren (ed.), *The history of statistics: Their development and progress in many countries* (New York, 1918), 576–577.

fledged instrument to monitor the overall state of American society," and thus began to fulfil the vision of mid-century practitioners of social statistics, who had seen it as "the centerpiece of any new effort" to "monitor, analyze and organize" the development of the American political and social economy.[46] Because population issues were what provoked Francis Walker's extensive, revealing and influential interventions in the spatial politics of government-ality, the lion's share of my attention will be directed at the population schedules in their connection to social policy. As I will treat it, then, govern-mentality was a discursive formation centered on the social and demographic dimensions of the US Census, both understood within a general political eco-nomic problematic. This discursive formation will be the topic of Chapters 2 and 3.

The American state, state theory and governmentality

A focus on national censuses brings me face to face with the question of how to understand the relationship between governmentality and the national state. The existing literature on governmentality reveals an ambivalent atti-tude toward the category of "the state." On the one hand, the term is used with great frequency, since no matter the scale at which its operation is being studied, the logic of governmentality presupposes some sort of governing "agency" or "subject," some actor or actors behaving according to its dictates. At any scale above that of individual (or perhaps family) government, "the state" is an obvious candidate, as the works cited in the genealogy given above attest. On the other hand, in keeping with Foucauldian analytical practice, there is a pervasive and principled distrust of any category as monolithic as "the state." As with "the human subject" in Foucault's earlier researches, "the state" is characterized at the level of theory as a contingent social construct, an "effect" lacking internal unity, clear boundaries separating it from its context, or autonomous causal powers. This ambivalence animates Foucault's seminal paper on governmentality, in which a long historical discussion of its logic, unintelligible apart from an assumption of some degree of efficacy on the part of national states, is followed by an injunction against paying too much attention to states: "But the state . . . does not have this unity, this indi-viduality, this rigorous functionality, nor, to speak frankly, this importance; maybe, after all, the state is no more than a composite reality and a mythicized abstraction, whose importance is a lot more limited than many of us think." The two contrasting emphases are brought together in the very next sentence: "Maybe what is really important for our modernity . . . is not so much the *étatisation* of society, as the 'governmentalization' of the state."[47] The

---

[46] M. Anderson, *The American census: A social history* (New Haven, CT, 1988), 85, 33.
[47] Foucault, "Governmentality," 103.

"governmentalization of the state" is precisely the way I would characterize the process I intend to chart in the context of the late nineteenth-century United States. But the phrase clearly implies some prior existence of the state, and implies also that the transformation wrought in the state has results that are interesting in some way. Making adequate sense of what occurred requires treating the state as more than merely an "effect." It requires supplementing nominalism with a more conventional willingness to allow the state provisional stability and efficacy where appropriate. Here my strategy will be to foreground the ambiguity detectable in Foucault's treatment of the state, and to interpret this ambivalence as an incipient rapprochement with more conventional analyses.

Earlier remarks Foucault made on the state in an interview given during the 1970s suggest that the ambivalence is not specific only to his work on governmentality:

I don't want to say that the State isn't important; what I want to say is that relations of power, and hence the analysis that must be made of them, necessarily extend beyond the limits of the State. In two senses: first of all because the State, for all the omnipotence of its apparatuses, is far from being able to occupy the whole field of actual power relations, and further, because the State can only operate on the basis of other, already existing power relations. The State is superstructural in relation to a whole series of power networks that invest the body, sexuality, the family, kinship, knowledge, technology and so forth. True, these networks stand in a conditioning-conditioned relationship to a kind of 'meta-power' which is structured essentially round a certain number of great prohibition functions; but this meta-power can only take hold and secure its footing where it is rooted in a whole series of multiple and indefinite power relations that supply the necessary basis for the great negative forms of power.[48]

The rhetoric of "negativity" signals that at this stage, Foucault had not yet formulated a clear idea of governmentality, but in so far as his view of the "valence" of the state would change, it would only become easier in principle to justify an analytical interest in the state as an effective agent. The state as a purely coercive power is not necessarily of much analytical interest, but a concept of the state which incorporates governmentality would greatly complicate (and hence render far more interesting) the question of the effects of state activity. The most important thing to note in this passage is the hesitant acknowledgment of a "conditioning-conditioned relationship" between structures of "meta-power" and the micro-power relations that invest them. This suggests that it is possible to view the relation between nominalistic contingency and provisional stability in a dialectical fashion.

To sum up my contention, Foucault implicitly accepted the idea that a social constructionism which fails to take into account the (at least transient) solid-

---

[48] M. Foucault, "Truth and power," in C. Gordon, ed., *Power/Knowledge: Selected interviews and other writings, 1972–1977* (New York 1980), 122.

ity of the products of social construction tells only half the story. He was not so interested in telling the more conventional half of the story, but he knew that there was some point in telling it. The state is derivative, not clearly bounded, internally fragmented and enjoys at best intermittent autonomy; but this "relative autonomy" is not utterly insignificant, and has arguably become more significant during the past century in many parts of the world.

Critical state theory offers a number of insights which come very close to a Foucauldian view in emphasizing the derivativeness, internal fragmentation and ambiguous boundaries of states.[49] But state theory has the great advantage of being better equipped to recognize and incorporate (rather than merely acknowledge in principle) inertia, solidity, and causal efficacy, whenever the processes through which states are constituted and reconstituted bring such characteristics into being. To make use of state theory is not automatically to surrender to an essentialist, monolithic view of the state. Bob Jessop's "strategic-relational" approach to state theory is particularly well suited to the rapprochement I would like to achieve here. This approach characterizes the state as a "form-determined condensation of the balance of political forces," which is to say that although the interests and programs animating state activity are largely of external ("social") origin, the state structures that have been left behind by previously important interests have a certain inertia to them, and thus end up inflecting (or in some cases deflecting) the projects mediated through them. State forms are constantly changing, too, but are not as liquid as the programs being pursued through them, and thus they exercise an unintentional "strategic selectivity" with regard to these programs. Therefore, "state forms have significant effects on the calculation of political interests and strategies and thus on the composition and dynamic of political forces. These forces may well attempt to use the state but neither they, nor it, can be seen as neutral transmission belts of interests which are fully determined elsewhere in society."[50]

In an interesting attempt to get to the bottom of, and then get beyond the seeming incompatibility between Foucauldian nominalism and Poulantzas's relational state theory, Jessop offers a provisional solution to the basic issue which seems eminently wise: "the diversity of micro-social relations is not without its own limits. For, although individual relations or institutions can be considered in isolation as polyvalent elements without any fixity, they are typically integrated into longer chains and systems of elements which restrict their fluidity and lability."[51] "The state" is an intelligible term for one such chain. As a "form determined condensation of the balance of political forces," it dialectically exercises "strategic selectivity" as a mediator of the discourses and practices flowing through it. The Gilded Age American state as a context

---

[49] *Ibid.*, 57, 61; B. Jessop, *State theory: Putting capitalist states in their place* (University Park, PA, 1990), ch. 12.   [50] *Ibid.*, 149.   [51] *Ibid.*, 242.

for early "governmentalization" can be characterized in these terms without siginificant loss of subtlety.

As long as the derivativeness of the general features of the state is kept in view, these features can help set the stage for the story told in the remainder of this study. By most measures (budget relative to size of the American economy, number of personnel relative to size of population or territory), the American federal state by the late 1870s was unusually weak and small as compared with the national states of other industrializing countries.[52] During and immediately after the Civil War, both the Union and the Confederate states had undergone vast expansions and strengthening of their powers to rule. The Union state, which was essentially an instrument of the Republican Party, arrogated to itself the means to enforce loyalty; used tariffs to protect indigenous industry; put the Union's financial system on a national footing in order to fund the war (through suspension of convertibility, the issuing of greenbacks, the cooptation of the nation's banks into a national system, and the creation of a major market in government securities); imposed drastic changes in property relations and citizenship through emancipation of slaves; established military rule and other aspects of reconstruction in the South; pursued territorial settlement through the homesteading program; and created a large class of stakeholders through the payment of veterans' pensions.[53]

By the end of Reconstruction, much of this leviathan had been dismantled (the veterans' pensions being the most interesting exception from a "governmentality" perspective), despite the hopes of a few that the Civil War had ushered in a new age of expanded state activity.[54] The reasons are complex, but a partial list would include the lack of a strong constituency for reconstruction other than southern blacks, the return of a two-party dynamic in Congress and sustained pressure from finance capitalists for retrenchment of government activity according to the principle of *laissez-faire*.[55] In addition to pensions, the chief material activities through which the state still made itself felt in national life after 1877 fell under three headings: (1) the fostering of national economic integration through the postal service, support for railroad construction and the system of tariffs; (2) the extraction of revenue through taxation of alcohol, tobacco and imports, which gave the federal state a noticeable presence in ports and commercial centers; and (3) military-led pacification of Indians and internal development and territorial expansion on the western frontier.[56] But scholars argue that these were modest measures, com-

---

[52] Mann, *Sources of social power*, vol. II, ch. 11.
[53] R. Bensel, *Yankee leviathan: The origins of central state authority in America, 1859–1877* (New York, 1990); T. Skocpol, *Protecting mothers and soldiers: The political origins of social policy in the United States* (Cambridge, MA, 1992), 63–151.
[54] Skocpol, *Protecting mothers and soldiers*; M. Keller, *Affairs of state: Public life in late nineteenth century America* (Cambridge, MA, 1977), 123.
[55] Keller, *Affairs of state*, 181; Bensel, *Yankee leviathan*.
[56] Bensel, *Yankee leviathan*, 400–402.

pared with the activities of European states. Overall, the American federal government had, according to Stephen Skowronek, returned to a mode of existence best understood as a "state of courts and parties."[57] Its structures had undergone shrinkage, and had once again become little more than institutional arenas for the relatively direct expression of competing party, sectional and other interests, all mediated through an active judiciary instead of large, stable executive bureaucracies. The patronage system, which Skowronek identifies as the most important single obstacle to the expansion of administrative capacity, once again animated the administrative apparatus and absorbed much of the time and energy of higher-level officials (see Chapter 2). Richard Bensel argues that it was only when the two major parties reached political parity after 1877 that the fledgling bureaucracies began to have opportunities to insulate themselves to some degree.[58] Once the Democrats could vie for real political dominance, the threat of faction within the Republican Party became a strong motivation for acquiescence to the demands of the Civil Service Reform movement.[59] Skowronek notes that as control of the executive began to see-saw between the two parties, it began to make sense for lame-duck presidents to secure their own appointees by expanding the list of Civil Service protected positions just before surrendering the administration to the opposing party.[60] Morton Keller stresses also the inhibiting effects of the localistic bias of American political sentiments and structures, most importantly, the constitutional principle of "devolution of powers," whereby all powers not specifically delegated to the federal government are the preserve of state and local governments.[61]

The only senses in which something like "governmentality" could be said to exist were in the system of pensions (since it created client relationships and mutual interests between the state and a large sub-group of the population), and in the judicial commitment to defend and strengthen the freedom of the internal market (though *laissez-faire* did not constitute a complete and sufficient governmental program in the eyes of any group other than domestic industrial and commercial interests). Governmentality in the narrower sense which I give it (see above) remained largely latent. The census was taken every ten years, but although it was beginning to be seen as a potential tool for national regulation, its administration was not immune from patronage. Furthermore, no permanent office yet existed to begin to give it some bureaucratic insulation from partisan politics. Other centers for the generation of national statistics were few, and no more advanced than the census in most respects. The "governmentalization of the state" in this context would involve

[57] S. Skowronek, *Building a new American state: The expansion of national administrative capacities, 1877–1920* (New York, 1982), 15. [58] Bensel, *Yankee leviathan*, 3.

[59] B. Silberman, *Cages of reason: The rise of the rational state in France, Japan, the United States and Great Britain* (Chicago, 1993), 256–259.

[60] Skowronek, *Building a new American state*, 47–84. [61] Keller, *Affairs of state.*

very specific programs: (1) the push for collection of more and better statistics; (2) the push to give these statistics a larger role in governmental decisionmaking; (3) the increasing involvement of social scientists in this decision-making; and (4) the establishment of institutional arrangements conducive to the pursuit of the first three goals. The most visible specific policy initiatives that represented incipient governmentalization of the late nineteenth-century American federal state were the push for an expanded census and a permanent Census Office, championed most effectively by Francis A. Walker; the establishment of the Bureau of Labor Statistics under Carroll D. Wright in 1885; certain aspects of the Civil Service reform movement; and the beginnings of regularized consultations between policy-makers and social scientists. All of these campaigns were embedded within a more extensive and diffuse discursive formation just taking shape at that time, and cannot be understood apart from it.

One of the basic claims I would like to make in this study is that "governmentality" is an analytically helpful concept, that it explains phenomena not so well explained by other approaches. Thus, I need to be as explicit as possible about its advantages, not only in relation to more abstract "isms" such as Marxist state theory (see above), but also in relation to more concrete perspectives taken in historical analyses of late nineteenth-century American state formation. The work of Skowronek, Bensel, Keller and others has been immensely fruitful, enriching and complicating the picture we have of the Gilded Age federal state. But despite the gains made, political histories of the period continue to miss governmentality, and thereby to miss some subtle features of the political culture. Although there is considerable variety in the work of these authors, it is fair to say that they have tended to focus more on measurable state power ("administrative capacity") than on philosophies or programs of government. This focus can be understood as a corrective to more traditional political histories, which tended to construct a two-stage sequence beginning with the Gilded Age (dominated by a *laissez-faire* philosophy) and moving rather quickly around the turn of the century to the Progressive Era (characterized by a sudden willingness to regulate economic activity). The great advantage of the notion of administrative capacity is that it shifts the question to the realm of what was actually there, institutionally speaking, in the way of state power. Most importantly, recent research has revealed considerable levels of state activity well before the Progressive Era, drawing attention particularly to the post-Civil War decade as a time of experiments in heightened federal power.[62]

Yet two features of this work tend to obscure the emergence of governmentality as a distinctive and potentially important logic. The first is the conviction that partisan politics severely circumscribed the sorts of reform efforts

---

[62] See in particular, Bensel, *Yankee leviathan*.

that would foster governmentality. Skowronek in particular argues that almost all efforts at Civil Service reform before 1900 were doomed to relative inconsequentiality by the patronage system.[63] In so doing, he leaves insufficient room for the sort of "dialectical" emergence of governmental thinking, represented by the likes of Francis Walker (see Chapter 3), into an initially hostile political culture. This oversight may also have to do with the insufficient attention paid by all of these authors to specifically statistical representations of social reality such as the census. I would suggest that the growing availability of detailed statistical information allowed patronage appointees like Walker to begin to transform themselves and their offices into havens for incipient modern expertise well before the grip of patronage was definitively loosened. I refer here not to what might be called (adapting a phrase from Poovey) "gestural governmentality," that is, declarations of support for the use of statistics or other empirical data in policy-making. Such declarations were common, but were rarely followed or accompanied by the actual use of statistics, which is what I mean by governmentality.

Secondly, the focus on administrative capacity obscures a distinction at the root of governmentality. The growth of administrative capacity is generally understood to involve two main processes: the (at least partial) insulation of a piece of the state apparatus from external (e.g., partisan) influence; and the organization of decision-making procedures and authority within the apparatus according to standard, logical rules.[64] These processes would facilitate the appearance of governmentality, but the notion of administrative capacity tends to conflate the capacity to regulate with actual regulation. The core logic of governmentality involves careful deliberation about whether and how much to regulate, a nuance missed by too exclusive a focus on things like the fiscal mass of the state or tallies of its personnel (see Chapter 7). The focus on party politics and on administrative capacity together predispose scholars to perpetuate the old binary "before and after" sequence in a new way, perhaps best captured by the headings under which Skowronek organizes his account: "state building as patchwork" and "state building as reconstitution."[65] However accurate these headings may be as characterizations of two eras in the growth of federal administrative capacity, they are insufficiently subtle to capture the notion of governmentality. And this blind spot makes it difficult for scholars to recognize governmental programs like that of Francis Walker, based as it was on a commitment to decide whether and how much to regulate, as anything other than transitional forms. A governmentality perspective allows us to look beyond administrative capacity and understand the role of empirically based decisions about whether to regulate as indications of a distinctive governing logic.

[63] Skowronek, *Building a new American state*, 47–84.
[64] Mann, *Sources of social power*, vol. II, 54–69.
[65] Skowronek, *Building a new American state*, v.

This brief characterization of the Gilded Age American state, and of what its governmentalization would require, should not offend Foucauldian sensibilities too strongly. In line with those sensibilities, the American state is here protrayed as derivative (during the Civil War and reconstruction, a mere shell for Republican Party initiatives, and afterwards almost wholly given over to the workings of patronage); to the extent that it was derivative, its boundaries were not clear (since many of its agencies were mainly vehicles for the expression of "external" interests); and it was fragmented (with various agencies, for example the army and the Bureau of Indian Affairs, frequently pursuing conflicting agendas). Yet this was an historical, not an eternal condition; over the next sixty years, precisely through the gradual consolidation of governmentality and bureaucratic rationality more generally, the American state would come to be less and less completely characterizable through Foucauldian methodological injunctions requiring an exclusive focus on micro-level strategies. To be sure, the federal government would never cease to be shot through with external programs and interests (both "micro" and "macro"); its boundaries would never cease to be fuzzy and permeable; and overall unity of purpose and action would continue to elude it (except to some extent during world wars). Nevertheless, its fiscal mass, the collection of individuals paid out of its budget, the fairly well-defined set of institutions in which they worked, the real property owned by "the federal government" as a legal entity, and the decisions emanating from this material, social and fiscal ensemble would all expand to play an ever greater role in American "society." The centerless thickets of micro-level power relations composing "state" and "society" would come increasingly to be circumscribed and organized by the durability of their creations.

I will not take my discussion of these theoretical issues any further for the present. Although these remarks are of course only schematic, they have perhaps at least lent theoretical plausibility to the undertaking. Again, to sum up, the ambivalent attitude toward the category of the state evident in Foucault's (and others') treatments of governmentality is a symptom of the fact that his precautions against taking monolithic categories for granted are just that: precautions, not absolute strictures. Foucault was unable to do without "traces" of "the state," and there is much to gain by more forthrightly acknowledging its usefulness as a category

Governmentality and geography

This leaves one last (and in some respects the most crucial) "context" issue: how might this great discursive apparatus bring about territorial mastery? Territory is a key concept both in Foucault's definition of governmentality and in Mann's definition of "infrastructural power," one of the analytical categories in state theory that approximates to at least some aspects of governmentality. Infrastructural power is " the institutional capacity of a central state . . . to penetrate its territories and logistically implement its decisions." It is, however, "a two-way street: it also enables civil society parties to control the state."[66] The notion of "fostering" central to governmentality is not well developed here, but the kind of two-way "penetration" with which Mann is concerned would require mutual recognition of interests. All types of social power network imply some specific organization of space. Since governmentality by definition involves "territory," and can only manifest itself as and through a set of material and human networks, it is no exception. In this final section, I will very briefly preview the geographical issues inherent in any program of national-scale governmentality. For analytical purposes, I will divide these issues into three parts corresponding with the three "moments" of the "cycle of social control" explained above (namely, observation, judgment and regulation or enforcement). Again, national scale governmentality, like discipline, involves all three of these moments.

In its observational moment (i.e., in the practice of census-taking), governmentality must make use of material networks. Thus some kind of rudimentary transportation and communication network is an indispensable spatial prerequisite. Beyond this, the possibility of acquiring comprehensive statistical knowledge of people and resources distributed over national territory depends on three geographical conditions, which I call abstraction, assortment and centralization. These three conditions together make it possible for subsequent acts of compilation to appear apolitical. Abstraction refers to the possibility of comprehensive epistemological access to all parts of the territory and everything within it. This brings into focus not only issues of physical infrastructure but also legal issues and the negotiation of privacy and the state's right to information. Assortment refers to the conditions that make it possible to pin down and distinguish different units of resources and especially people, to make them susceptible to enumeration. This leads into political issues of residential fixity and family structure. Finally, the quality and usefulness of the knowledge gained through censuses depends in part on the centralized organization of knowledge gathering. This in turn depends not only on physical infrastructure but also on other conditions that need to be in place to ensure that information is faithfully recorded at the point of collection of knowledge. In concrete terms, centralization of knowledge gathering immersed late nineteenth-century census takers in the political morass of patronage and its links with associated localism. When governmental knowledge gathering is based on sufficiently effective abstraction, assortment and centralization, it is much easier for authorities to convince the public that information is properly handled once it arrives at "centers of calculation" (to borrow a term from Latour). To summarize, the observational moment of governmentality

---

[66] Mann, *Sources of social power*, vol. II, 59.

required epistemological mastery of national territory. However, all three elements of epistemological mastery pivot on a tension between fixity and mobility. This epistemological geography of governmental power will be the subject of Chapter 5.

I will argue in Chapter 6 that a spatial politics of territorial mastery inflects the second "moment" in the cycle of social control ("normalizing judgment") as well. Here, the focus will be on Francis Walker's political-economic understanding of American society. This political economy constitutes the lens through which he made sense of the social economy constructed by census results. Any understanding of a national social body at least implicitly assumes a set of spatial ordering principles. If a collection of people, activities and resources in a territory is to be conceived as a larger unity of some sort, (1) some sense must be made of regional differentiations, complementarities and interrelationships (in economic, cultural or demographic terms as well as in terms of resource endowments); (2) existing relations of concentration and dispersion (relations of urban to rural life) must be interpreted; and (3) there must be some at least implicit notion of how fixity and mobility should be distributed among different groups of people, activities and resources. In short, every conception of proper social order, when brought into relation with empirical knowledge produced by censuses, etc., implies a proper spatial order. As with the moment of observation, the moment of (political economic) judgment will also turn out to revolve crucially around a tension between fixity and mobility.

In the final substantive chapter (7), I will focus on space as a tool of material social control, on regulation of the location and movement of people and resources through national space. Regulation is the "last" moment of the cycle of social control in both conceptual and historical terms. In the United States, many significant advances in the regulatory mastery of national territory awaited the twentieth century. This was only partly due to the cautious logic of governmentality whereby it is often prudent not to regulate. It also had much to do with the forces arrayed against major expansions in the power of the national state (the tenacious hold of patronage, a pervasive localism, see above). The practical result was that sensitively calibrated regulation of movement and policing of fixity within national territory was seldom feasible. Thus some of the first effective interventions that can be understood as expressions of national-scale "governmentality" occurred at the borders of the territory.

Part I

To ask how a discourse of governmentality emerged in late nineteenth-century America is to ask familiar "who," "what," "when" and "where" questions. Partly in order to ensure a cautious approach, but mainly because it offers such marvelously sensitive guidelines for historical description, I will tell this story of emergence with the help of Foucault's "archaeological" method. Although there are obvious affinities between his earlier and his later studies, there is no rigid requirement that anyone studying governmental discourses must use archaeological method. Archaeology will be put to work here as a systematic and yet non-totalizing set of organizing concepts on which to hang a specific narrative. Again, the archaeology can be seen as an instance of what Foucault terms "general history," that is, history devoted to laying out "spaces of dispersion" within which concrete events can be described with the greatest possible specificity.[1] Rather than anchor historical narratives in *a priori* "unities" (*oeuvre*, *Zeitgeist*, "period," "progress") which constitute so many temptations to simplification and reduction, Foucault argues for "the project of a *pure description of discursive events* as the horizon for the search for the unities that form within it" (italic in original).[2]

Because I am using the archaeological method as a subordinate tool in a larger, not strictly archaeological undertaking, I will not wipe the interpretive slate quite so clean as Foucault would have it wiped. Nevertheless, the particular way in which Foucault pursued this elusive "horizon" was immensely interesting, and will remain very useful as a regulative ideal. My approach here illustrates the commitment (discussed in the Introduction) to avoid excessive timidity in using Foucault's ideas where they seem appropriate. If this practice seems "uncritical," it is because I have run across not a single convincing critique of Foucault's archaeological method. Most challenges to Foucault's archaeology are aimed at his general structuralist demotion of the human subject as an explanatory category. I think such critiques are misconceived in much the same way as is the claim that a Foucauldian analysis cannot make

---

[1] M. Foucault, *The Archaeology of knowledge* (New York, 1972), 10.   [2] *Ibid.*, 27.

use of the category of "the state." In both cases, Foucault's insistence on the *derivative* nature of a category is misunderstood as a claim that the category has no effectivity and thus no role in social causation.

The purpose here and in Chapter 2 is to describe the discursive formation of governmentality as it emerged in late nineteenth-century America. A "discursive formation" is a "positivity" of discourse, a set of concrete discursive practices that have a very specific kind of unity. "To analyze positivities is to show in accordance with which rules a discursive practice may form groups of objects, enunciations, concepts or theoretical choices."[3] In other words, discursive formations are not identifiable on the basis of shared objects, bodies of knowledge, styles of argumentation, key categories or themes, but rather on sets of implicit protocols for how such things can be recognized as legitimate or acceptable in the first place. Many of the discursive formations Foucault analyzed in his early studies played key roles in constituting some of the human and biological sciences (psychiatry, evolutionary biology, clinical medicine, philology), and so eventually acquired considerable unity and explicitness in their rules of formation. Governmentality has never achieved such "tightness." It has always been something more and something looser than "political economy" or "sociology" or any other neatly identifiable discipline or institutional site (see Chapter 1). Thus its rules of formation have always been frayed, contestible and thoroughly entangled with "external" developments.

Nevertheless, the basic "dimensions" along which Foucault suggests discursive formations should be analyzed are an excellent way to organize and locate governmentality as it appeared haltingly and unevenly in late nineteenth-century American public life. In Chapter 2, I will focus on the first dimension: the formation of objects. The second dimension, the formation of "enunciative modalities" (or "subject positions") will be the focus of Chapter 3. The guidelines Foucault provides for studying these two aspects of discursive formations constitute the most useful and interesting part of the archaeological method. In Chapters 4 and 6, I will address features of governmental discourse that would fall under the last two "dimensions" (formation of concepts and formation of theoretical strategies), but no longer from an explicitly "archaeological" perspective. This is simply because I believe there is much less in the way of historical and geographical insight to be gained from continuing to apply archaeological method throughout. The attempt would render these chapters considerably more cumbersome but would not yield corresponding analytical rewards.

Once again, the objects of national-scale governmentality include the population and its activities, as well as non-human resources, distributed over a clearly delimited national territory, plus the relations between all of these. How did all these things come to be linked together conceptually by contemporaries and understood as components of the same overall complex?

---

[3] *Ibid.*, 181.

2

The formation of governmental objects in late nineteenth-century American discourse

In Chapter 1 the discourse of governmentality was defined as having two levels: the *connaissance* of political economy and the *savoir* of social economy. If we follow Procacci in conceiving of the two levels as distinct but interdependent, it becomes clear that the formation of governmental objects takes place largely in the realm of social economy. Foucault distinguishes three sub-questions that need to be answered: (1) what were the "surfaces of emergence" of governmental objects, the concrete social and institutional "sites" at which they became visible? (2) What were the "authorities of delimitation" of the objects, the individual or collective agents generally recognized as capable of delimiting and designating the objects of governmentality? Lastly (3), what were the "grids of specification," the "systems according to which the different 'kinds of [object]' are divided, contrasted, related, regrouped, classified, derived from one another as objects"?[1]

Surfaces of emergence

I will leave the question of how census information was produced until Chapter 5 (where it receives an extended treatment), and ask here only how and where it appeared once produced. To understand what Foucault means by "surface of emergence," it may help to refer to the examples he gives from his own research. In the *Archaeology*, he uses the nineteenth-century discursive formation surrounding the object "madness" to illustrate this term. Although madness was a condition attributed to individuals, it concretely appeared, became visible at specific socially defined locations: "the family, the immediate social group, the work situation, the religious community," and increasingly also in art, discourses of sexuality and penality.[2] The situation here is quite different: in a sense, the "surface of emergence" of the governmental

---

[1] G. Procacci, "Social economy and the government of poverty," in G. Burchell, C. Gordon and P. Miller (eds.), *The Foucault effect: Studies in governmentality* (Chicago, 1991), 151–168. M. Foucault, *The archaeology of knowledge* (New York, 1972), 41–42.   [2] *Ibid.*, 41.

objects is already given in the census itself. But the census did not simply and effortlessly appear before the eyes of the nation. On the contrary, making the statistically constituted national social body available for the inspection of government officials and the interested public was a daunting task in the late nineteenth century. In order fully to understand the problem, we must now plunge into the thicket of historical detail. Following the dissemination of the results of the Tenth (1880) Census, we can trace its surfaces of emergence and appreciate the difficulty of the undertaking.

The Census of 1880 was to be a particularly important enumeration, since it was to mark and document the nation's first century of existence. Thus Congress provided for an unusually complete publication. A full set of the results comprised twenty-two quarto volumes, and the shorter *Compendium* two quarto volumes. Once compiled, the results of each decennial census (starting with the Ninth in 1870) were to be printed and bound by the Government Printing Office and distributed by the superintendent of documents in the Interior Department.[3] An 1882 Act of Congress provided for the publication of the "usual number" of copies of each (at that time, the "usual number" was set at 1,900 for most Congressional documents) for distribution to federal officials, members of Congress and designated libraries, plus 100,000 additional copies of the *Compendium* and 10,000 additional complete sets of results for wider distribution. The volumes in the complete set on population and agriculture were to be printed in 20,000 copies each rather than 10,000, and the expected cost of all this (plus a few special reports) was $835,461.61.[4] Because of delays in checking proofs of the compiled results, the GPO did not begin to publish them until 1883. The public printer, in his 1883 Annual Report to Congress, marvelled at the sheer scale of the undertaking: "At times, this office has had as high as *twenty tons* of long primer, brevier and nonpariel type, rule and figure work, locked up, awaiting return of proofs; probably the largest amount of 'live matter' ever kept standing at one time in this or any other country."[5] An 1884 act called for an additional 25,000 copies of the *Compendium*, as well as various special reports in much smaller runs of 2,000–3,000 copies each, and later acts added requests for an additional 10,000 copies each of the population volume and the agriculture volume. All told, 615,500 copies of volumes reporting the results of the Tenth Census were distributed, but the printing was not

[3] L. Schmeckebier, *The Government Printing Office: Its history, activities and organization* (Baltimore, MD, 1925), 14, 31.

[4] US Congress, "An Act to provide for the publication of the Tenth Census," Aug. 7, 1882, *US Statutes at Large* vol. 22 (47th C.), 1882, 344–345. Congress had ordered the *Compendium* of the Seventh (1850) Census published in 320,000 copies, but this still represented a far smaller, more manageable publishing job than that of the Tenth Census (see M. Anderson, *The American census: A social history* (New Haven, CT, 1988), 53).

[5] Government Printing Office, *Annual Report of the Public Printer*, Senate Misc. Documents 10 (48th C., 1st S.), US Serial Set 2170 (Washington, DC, 1883), 14.

finished until 1888, and the distribution was not completed until the early 1890s.[6]

The basic means by which the distribution of documents achieved reasonable geographical coverage of the nation had evolved only slowly during the nineteenth century. From 1813 until 1857, the secretary of state had taken responsibility for carrying out an Act of Congress requiring that one copy of each journal and document be sent to each university, college and historical society incorporated within each state.[7] This did not involve a large number of institutions, and it is difficult to ascertain how many additional copies of decennial censuses were actually sent out. Laws in force since the late 1850s had assigned responsibility for nationwide dissemination to a superintendent of documents under the secretary of the interior, and had established the privilege of each senator, representative and territorial delegate to select one receiving library in his state, district or territory. But all of this did not constitute much of an improvement. The first historian of the GPO could still lament as late as 1882 that "no proper provision has been made for the dissemination of these costly publications."[8] Of the 1,900 copies comprising "the usual number," approximately 900 were normally distributed within the federal government. The bulk of the remaining 1,000 or so were to be distributed to the Senate and House Document Rooms and to the Interior Department (as outlined above).[9] An 1895 Act added state and territorial government libraries to the recipient list, and subsequent Acts in the early years of the twentieth century further expanded it.[10] One of the earliest printed lists of designated libraries appears in an 1886 letter from the secretary of the interior to Congress. The total number of libraries designated therein is 385, distributed among the states according to the size of their congressional representations. The number of libraries in each state ranged from thirty-three (New York) and twenty-seven (Pennsylvania) at the high end to one in each of the most sparsely settled territories. The bulk of recipients were town or city public libraries, college and university libraries and local library

---

6 C. D. Wright and W. Hunt, *History and growth of the United States Census*, Senate Document 194 (56th C., 1st S.), US Serial Set 3856 (Washington, DC, 1900), 68; US Congress, "An Act making appropriations for sundry civil expenses of the Government," July 7, 1884, *US Statutes at Large* vol. 23 (48th C.), 212–213; "Letter from the Secretary of the Interior transmitting a report by the Superintendent of Documents regarding receipt, distribution and sale of public documents on behalf of the Government," Feb. 18, 1890, House Executive Documents 212 (51st C., 1st S.) (Washington, DC, 1890).

7 Schmeckebier, *The Government Printing Office*, 31.

8 R. W. Kerr, *History of the Government Printing Office (at Washington DC), with a brief record of the public printing for a century, 1789–1881* (New York, 1970 [1882]), 62.

9 Government Printing Office, *Annual Report of the Public Printer*, Senate Misc. Documents 12 (47th C., 2nd S.), US Serial Set 2083 (Washington, DC, 1882), 7.

10 Schmeckebier, *The Government Printing Office*, 14, 31–32. Interestingly, it was not until 1913 that existing depository libraries were given permanent status and thereby protected from Congressional caprice.

associations. State historical societies also received copies, as did a few high schools and academies.[11] The other regular distribution outlets were congressional: the two document rooms were authorized to sell copies of documents to the public, and the representatives and senators had long taken advantage of franking privileges to send copies out to constituents from the Document Room stocks. Finally, citizens could write to the GPO requesting documents with advance payment, but there was no systematic way for citizens to learn what was to be published, or precisely what had been published, until the early 1890s.[12]

Since the Tenth Census was published in so many more than the "usual number" of copies, the existing system of distribution (involving a scant 400 or so regular recipients nationwide) was plainly inadequate. An 1882 report of the House Committee on Printing recommended temporary adoption of an extended depository system, with each senator choosing fifteen receiving institutions and each representative ten.[13] This plan was passed into law the same year, along with a requirement that the superintendent of documents in the Interior Department furnish a complete list of recipients.[14] It was not until 1890 that the secretary of the interior was able to produce even an incomplete version of this expanded list. At that time, roughly 80 percent of the distribution of census results had been completed, and the recipient institutions numbered in the thousands. The general pattern revealed in this list seems a fairly straightforward reflection of the central place structures of the various states at the time. The primate and second-tier cities boasted many recipients and the lower level towns fewer. But in states such as New York and Pennsylvania with relatively high population densities, copies of the Tenth Census appear to have reached down to quite local levels of the hierarchy. The 223 institutions receiving copies in New York state represented every county, and few interested citizens would have been more than a few miles from a copy.[15] Even this relatively comprehensive coverage fails to account for all of the 10,000 complete sets of results, and the fate of most of the 100,000 *Compendia* must remain a matter of conjecture. In the best possible scenario, the Tenth Census would truly have blanketed the land.

Yet even had this been the case, there was another important consideration in the emergence of statistical governmental objects: time. The kind of national-scale governmentality with which this study is concerned does not in

---

[11] "Letter from the Secretary of the Interior submitting a report of the receipt and distribution of public documents on behalf of the Government by the Department of the Interior," Feb. 17, 1886, House Executive Documents 78 (49th C., 1st S.), US Serial Set 2398 (Washington, DC, 1886), 2–8.   [12] Kerr, *History of the Government Printing Office*, 62, 174.

[13] US Congress, *Report of the House Committee on Printing*, House Reports 1784 (47th C., 1st S.), US Serial Set 2151 (Washington, DC, 1882).

[14] US Congress, "An act to provide for the publication of the Tenth Census."

[15] "Letter from the Secretary of the Interior transmitting a report by the Superintendent of Documents," 14–37.

general depend crucially on speed; the decennial census appeared then in a rhythm best suited to very general, "macro" level social policy making. Nevertheless, there were important considerations favoring timely dissemination of Census results. The most obvious was Congressional reapportionment, and thus the superintendent of the Tenth Census concentrated on finalizing population numbers first. By late 1881, Superintendent Francis Walker was able to lay out for Congress a range of alternative reapportionment scenarios based on different total numbers of representatives (the total size of the House increased steadily throughout the nineteenth century, and stood at 293 in 1880).[16] The actual reapportionment was carried out in February of 1882.[17] The volume in the complete set of results concerning population was the first of the twenty-two volumes to be printed in 1883. The most important general circumstance demanding speed of publication was the unusually disruptive social and economic change American society was undergoing at the time. Although almost any period of American history can be seen to have involved rapid change, it is fair to say that late nineteenth-century industrialization, immigration, urbanization and the emergence of a modern middle class together constituted a particularly disorienting experience.[18] For social reformers and early social scientists, the already considerable level of general interest in social statistics would have acquired extra urgency in so far as the numbers were coming to be seen as tools of social regulation. Statistics even a few years out of date could be quite misleading.

Until the Civil War era, the most important "timely" surface of emergence for statistically constituted national objects of governmentality was a small group of merchant magazines. Foremost among these were *Niles Weekly Register* (established in Baltimore in 1811), *Hunt's Merchants' Magazine and Statistical Review* (New York City, from 1839) and *DeBow's Review* (New Orleans, from 1846).[19] These magazines originated mainly as informants of merchants, and as outlets for the trade statistics compiled by the federal government (the motto adopted by J. D. B. DeBow, for example, was "Commerce is King"). With this emphasis, it is not surprising that they appeared in coastal mercantile cities. Their circulations were quite small (none of the three sustained circulations much over 5,000 at mid-century).[20] Yet at

---

[16] Office of the Census, *Annual Report of the Superintendent of the Census*, House Executive Documents 1 (47th C., 1st S.), (Washington, DC, 1881), 665–727.

[17] Wright and Hunt, *History and growth*, 921.

[18] R. Wiebe, *The search for order, 1877–1920* (New York, 1967).

[19] R. C. Davis, "The beginnings of American social research," in G. Daniels (ed.), *Nineteenth century American science: A reappraisal* (Evanston, IL, 1972), 158; P. C. Cohen, *A calculating people: The spread of numeracy in early America* (Chicago, 1982), 165.

[20] D. Haury, "*Niles Weekly Register*," and D. Kohut, "*DeBow's Review*," both in A. Nourie and B. Nourie (eds.), *American mass-market magazines* (Westport, CT, 1990), 95–97 330–332; F. L. Mott, *A history of American magazines*, 5 vols. (New York, NY, 1930 [vol. I]; Cambridge, MA, 1938 [vols. II, III]; 1957 [vols. IV, V]), vol.1, 269, 697–698; vol. II, 339, 348.

the outset of the post-Civil War period, these journals were often the first places interested citizens and even government officials could find extensive reporting of census results.[21] Despite their meager distribution they were more easily available than results obtained through the vast but cumbersome and haphazard Federal Government distribution.

These magazines declined quickly after the Civil War in the face of competition from newspapers, which became the surfaces of most rapid emergence for governmental objects. The telegraph was the key catalyst for this change. Before the Civil War, newspapers were mainly local, both in their gathering and in their distribution of news; distant news came in, but only slowly. From the early 1850s, the expanding telegraph system allowed a New York City news cooperative to expand into the first national wire service, and the Associated Press soon handled the bulk of trans-local news gathering and redistribution (at least at the upper levels of the urban hierarchy).[22] The practice of retaining individual correspondents in centers of news creation became widespread during the 1850s.[23] Thus any census results that appeared could quickly be passed on to editors waiting at any of the 12,510 telegraph offices counted in the Tenth Census, as could any advance indications from Census Office officials. By the postwar period, superintendents in Washington had established a regular practice of issuing bulletins giving preliminary results and approximate totals to news services long before the GPO published final copies for distribution. These bulletins, not the final published volumes, became the standard source for newspapers. Their rhythm of appearance gives a quite accurate tracking of the ongoing compilation work in the Census Bureau offices. Since the bulletins were constructed for general consumption, every newspaper had the same information, and the reporting of any one paper can serve as an rough guide. For example, the *New York Times Index* reveals that estimates of population totals for the states of the union and for many cities were available by late 1880. By early 1881, further reports (and firmer tallies) of population numbers were interspersed with early results on fisheries; as the year wore on, bulletins began to appear detailing agricultural results and more specialized or derivative population figures, as well as mining, manufacturing, land ownership, municipal debts and bondholding.[24] All of these topics were presented to the public through newspapers far in advance of their appearance in bound volumes. But newspaper reporting of census results was necessarily very selective. Space allowed only aggregate summaries of the already abbreviated bulletins, or highlights of local or topical interest. Typical articles compared population or agricultural totals for the several states, or perhaps the

---

[21] Haury, "*Niles Weekly Register*," 332.
[22] A. M. Lee, *The daily newspaper in America: The evolution of a social instrument* (New York, 1973), 493–494, 500–502.   [23] *Ibid.*, 505.
[24] *New York Times Index: A book of record, 1880–1885* (New York, 1966), 115–116, 202, 275, 297.

largest cities in the nation, sometimes noting rates of change since the 1870 Census.[25]

To sum up, there was a tradeoff between the depth offered by the sluggishly emerging full sets of results and the timeliness offered by highly selective newspaper reporting. Any serious policy-making, even the quite urgent task of reapportioning seats in the House, had to await the cross-checked final results (or something very close to final). The urgent but less consequential public curiosity could be gratified at the earliest opportunity without undue concern for ultimate accuracy. Both surfaces of emergence were geographically comprehensive in their distribution; they could present readers almost anywhere in the national territory with the objects of governmentality. The vast majority of citizens who had a chance to observe the statistically constituted nation undoubtedly did so through newspaper reporting, and I suggest that the constraints of selectivity imposed on this form of reporting would have tended to encourage a sense of "nationhood." Due to lack of space, most census-related newspaper articles dealt mainly in aggregate national and state-level figures, and thus helped to reify not only individual states but also the nation as a whole (see Chapter 5 for a more extended discussion of the reification of the nation and its individual states).

Authorities of delimitation

The second sub-question one would ask in exploring the formation of objects of a discourse regards what Foucault calls "authorities of delimitation." Who can legitimately define, delimit or indicate the objects that emerge on the surfaces detailed above? In one sense, this is not such a complicated question here as it is in the case of early psychiatric discourse, for example.[26] The objects defined by the US Census are delimited first in principle by Congressional legislation and then in practice by the Census Office, which is headed by the superintendent of the census. But the question of Congressional control and of precisely who is in charge of the Census Office (and how that person is selected) turn out to have been very important in this phase of census history.

As noted in Chapter 1, the Gilded Age was one in which partisan politics reigned supreme over national administration through the instrumentality of party patronage. The Census Office was no exception. Not only did Congress legislate in great detail the terms under which the censuses would be taken, it also had the power not to legislate, and the power not to fund. Congress used all of these tools to circumscribe the possibilities for an "apolitical" enumeration. Legislation covered every aspect of the logistics of enumeration, as well

---

[25] *New York Times*, Jan. 1, 1881, 4; Nov. 8, 1881, 4; July 30, 1882, 6; Aug. 21, 1882, 4.
[26] Foucault, *Archaeology*, 41–42.

as the numbers and pay of employees; perhaps most importantly, it specified the structure and details of the actual schedules, the "grids of specification" against which the objects of governmentality would become visible.[27] There was some minor wiggle-room in the 1879 legislation allowing the Superintendent to formulate special schedules as needed for some of the economic statistics, and it appears that he was able to modify the format of the schedules somewhat without any specific Congressional approval.[28] Nevertheless, Congress fundamentally decided what would be asked.

Thus the most a superintendent could hope for in setting the structure of the census was to have been previously involved in the formulation of the census bill, as Francis Walker had been in 1869 and 1879. After all, however much members of Congress insisted on having the final say, most realized that they required assistance in writing Census bills that would at least be credible from a disinterested perspective. However, even so fortunate a circumstance did not necessarily result in increased influence on the part of the superintendent. The 1869 Census bill which Walker had helped to draft and which James Garfield introduced into Congress would have transformed the 1870 Census into something much more like its successors (much more modern and much more directly controlled by the superintendent). This bill met strong resistance partly because of suspicions about its electoral implications, partly because its elaborate character was seen by some as extravagant, and partly because it would have switched control of the patronage appointment of enumerators from the Senate to the House. Even after being reduced to a pale shadow of its initial form, little more than an adjustment of previous administrative procedure, the bill was tabled, and the 1870 count was taken under the 1850 legislation.[29] The bill for the 1880 count, again written by Garfield and Walker, was more skillfully designed, and made it through Congress in January of 1879. But Walker's insistence on a rudimentary merit examination in hiring the necessary army of enumerators, and his general refusal to bow to Congressional patronage requests, led Congress to refuse him supplementary funding when previous appropriations ran out in 1881. He resigned to take over the presidency of MIT, leaving charge of the remaining compilation to his assistant Charles W. Seaton.[30]

These episodes illustrate the ease with which Congress could use legislation and the purse strings to delimit the authority of the superintendent. Superintendents were better able to leave their mark in the compilation and reporting stages, when their work was somewhat better sheltered from partisan oversight. Superintendents were hemmed in to a great extent by the need

---

[27] US Congress, "An act providing for the taking of the Seventh and subsequent Censuses," May 23, 1850, *US Statutes at Large, v. 9* (31st C., 1st S.), ch. 11, 428–436; "An act to provide for taking the Tenth and subsequent Censuses," Mar. 3, 1879, *US Statutes at Large, v. 20* (45th C., 3rd S.), ch. 195, 473–481.   [28] Wright and Hunt, *History and growth*, 54.

[29] Anderson, *The American census*, 76–78.   [30] *Ibid.*, 95–100.

to compile statistics in a manner that allowed longitudinal comparison with the results of earlier censuses. But they could organize the published volumes, and issue news bulletins, so as highlight particular results as important. They could also do a great deal to affect public perceptions of the "objectivity" of the census by the attitude they took toward the results in their public remarks and writings. Francis Walker is also notable in this regard. One of his most significant decisions as superintendent of the Ninth (1870) Census was to criticize its results himself.[31] This was not a purely noble expression of disinterested judgment. As we have seen, the law under which he directed the enumeration was the 1850 law, not the one he and Garfield had taken so much time to craft. His criticisms were therefore not really attacks on his own work. Nevertheless, his willingness to point out faults in an enterprise that had absorbed so much of his energy over a span of years made a favorable public impression. My point in bringing this up is to suggest that despite the overweening presence of Congress in census matters, who the superintendent was mattered quite a lot during this period of American history. Before the turn of the century, the balance between the two authorities of delimitation (Congress and the superintendent) depended very much on the particular agendas and qualities of both. For all intents and purposes, the superintendency would be the vehicle for the emergence of governmental thinking about the census in the national state.

In order to place selection of the superintendent in proper context, it is necessary to revisit controversies surrounding pre-Gilded Age censuses, and also to review the basic outlines of the Civil Service reform movement in the Gilded Age. A number of scholars agree that the controversy surrounding the Sixth (1840) enumeration was a watershed in census history.[32] The scandal had to do with statistics of black insanity. William Weaver, who compiled and published the results of the Sixth Census, was accused of tampering with the numbers after prominent early statisticians exposed a large discrepancy between reported totals of insane blacks in northern states and the actual census returns. As compiled by Weaver, it appeared that rates of insanity among blacks increased drastically from South to North, and southern apologists for slavery predictably pounced on this as evidence for the beneficent character of the "peculiar institution." When the groundlessness of the apparent pattern was exposed, there was a great public outcry. Patricia Cohen has called this episode the "end of innocence" in the American love affair with numbers.[33] It neatly illustrates two things easily taken for granted in later years: (1) that measurement schemes require the disciplining of the measuring subjects (authorities of delimitation), as well as of the objects

---

[31] Office of the Census, *Ninth Census: The statistics of the population of the United States*, vol. .I (Washington, DC, 1872), ix–xliv.
[32] Anderson, *The American census*, 29–31; Cohen, *A calculating people*, 204; Davis, "The beginnings of American social research," 162.   [33] Cohen, *A calculating people*, 204.

themselves; and (2) that this involves the subordination of personal and political interests.[34]

From the 1840s onward, the question of who should be put in charge of the census came to be understood in the terms provided by a larger movement to establish integrity *and* competence as the characteristics necessary for public service in government posts. The Civil Service reform movement gathered steam especially in the decades following the Civil War. Its central target was the system of patronage. In this system, whichever of the two national political parties was in power could expect the president and members of Congress to reward its electoral campaign efforts by appointing thousands of party loyalists to Federal posts. Since the 1830s, patronage had been justified on the grounds that the greatest threat to honest pubic service was a long term in office.[35] According to this logic, patronage appointees, since they were members of the party in power, could at least be counted on to act in rough agreement with the "will of the people" as this was expressed in election results. The parties benefited through regular monetary assessments collected from patronage appointees as finders' fees for their jobs. This money was then used to finance future congressional and presidential campaigns in the hope of perpetuating party dominance in national government.

A number of circumstances conspired to place this system in a more unflattering light after the Civil War.[36] Among these was a degree of genuine public outrage over the way some patronage appointees used their appointments as vehicles for personal enrichment. The most spectacular example of this was the post of collector of customs at New York City. During the 1870s, more than 50 percent of total Federal revenue derived from duties and penalties collected at this one port, and a smoothly functioning arrangement was in place allowing officials to siphon off some of this flow.[37] The system of "moieties" had been authorized by Congress in 1789, and remained in place through most of the nineteenth century. According to this system, one half of each fine or forfeiture assessed by the collector or his agents went to the US Treasury. One quarter went to the informer who had alerted the collector, and the remaining quarter was divided among the collector, the naval officer in charge at the port, and the "surveyor." This arrangement had been intended to encourage zeal in inspecting goods, but, not surprisingly, it quickly became a source of vast wealth for the officials and their henchmen. As New York collector during the 1870s, Chester A. Arthur reported an official, legal annual

---

[34] T. Porter, *Trust in numbers: The pursuit of objectivity in science and public life* (Princeton, NJ, 1995), 22, 74.

[35] B. Silberman, *Cages of reason: The rise of the rational state in France, Japan, the United States, and Great Britain* (Chicago 1993), 234–242.

[36] *Ibid.*, 250–268; S. Skowronek, *Building a new American state: The expansion of national administrative capacities, 1877–1920* (New York, 1982), 39–59.

[37] Skowronek, *Building a new American state*, 61; L. White, *The Republican era, 1869–1901: A study in administrative history* (New York, 1958), 118–123.

income of approximately $40,000.00.[38] It was widely known that, in addition, bribes and unofficial assessments were endemic at all levels in New York, Philadelphia and New Orleans. The New York collector's post was the jewel in the national patronage crown, but major opportunities for corruption also surrounded the activities of the Interior Department, especially the agencies within it through which major funds flowed (the Bureau of Indian Affairs, the Pension Office and the Bureau of Lands).[39]

The reformers fighting this system were a small and self-concious group of elite northeastern (mostly New England) Republicans who were dissatisfied with the increasingly marginal position with respect to national power.[40] They perceived a need for more modern, efficient national administration, and felt that members of the educated elite were best qualified to construct and staff it. The best-known reformers were Carl Schurz, who became secretary of the interior in 1876 under Hayes, E. L. Godkin, editor of the *Nation* and George W. Curtis, who would be appointed to head the Civil Service Commission.[41] The typical supporter of their cause would have been a member of an old New England family and a graduate of Harvard, Yale, Amherst College or one of the other colleges of the New England elite. He would be a doctor, lawyer, clergyman or academic. He would in all likelihood have had personal experience or close family in the abolitionist movement leading up to the Civil War, and perhaps have served in the Union army as an officer.

The first vehicle of publicity for this group were the wartime reports of the United States Sanitary Commission during the Civil War. Subsequently they waged their campaign in the pages of a few distinguished national magazines: the *North American Review, The Forum, Harpers', The Nation, Atlantic Monthly* and *Century*.[42] After the war, the reformers came together under the umbrella of the newly formed American Social Science Association (ASSA), and from there continued their campaign without pause.[43] They made little headway in practical terms before the 1880s, but could at least balance their critiques by pointing to a few exemplary officials already at work in the federal trenches during the 1870s. Writing in the 1950s, historian Leonard White echoes what must have been the sentiments of many reformers at the time in his paean to these intrepid officials:

[I]n the midst of these scenes of betrayal [of the public trust] labored some of the ablest and most trustworthy public officials in the history of the Republic: men like Hamilton Fish, Secretary of State (1869–1877); Benjamin Bristow, Secretary of the Treasury (1874–1876); Jacob D. Cox and Carl Schurz, Secretaries of the Interior (1869–1870; 1877–1881). Nor should these names obscure those of secondary rank of equal integrity and ability – men such as William Hunter and A.A. Adee, whose collective

---

[38] White, *The Republican era*, 118, footnote 24.   [39] *Ibid.*, 180.
[40] Skowronek, *Building a new American state*, 43–45; Silberman, *Cages of reason*, 252–253.
[41] Skowronek, *Building a new American state*, 49, 52.   [42] *Ibid.*, 44, 53.
[43] Silberman, *Cages of reason*, 253.

service in the State Department . . . spanned nearly a century; Sumner I. Kimball, Superintendent of the Lifesaving Service; and Francis A. Walker, who created the modern foundations of the Census.[44]

The Census Office was associated with patronage corruption only in minor ways. Until 1900 it did not even exist as a permanent office; and when it did exist, it was not a conduit for massive flows of funds. Yet staffing the Census Office was a matter of political appointment, and even after the Pendleton Act of 1883, hiring there was exempted from regulation by the Civil Service Commission. As patronage positions dwindled in other agencies, Census Office hirings became the focus of ever more intense patronage pressure.[45] This automatically raised the standard questions about integrity and competence. As the scandal of the 1840s showed, integrity in the central office was a prerequisite for an "objective" census, but the important threats to integrity had less to do with greed than with the politically motivated willingness to distort results. Competence (both organizational and statistical) on the part of the superintendent was crucial. The logistics of the enumeration itself were daunting, and were always made more difficult by inadequate funding. In addition, the orderly digestion and cross-checking of incoming returns demanded skilled judgments about the quality of the information collected, a scrupulous approach to the mathematical operations performed upon it, and above all, patience. As noted above, some of the delays in publishing census results derived from the requirement for thoroughness and independent checking in compilation. Although the age of the professional statistician was yet to come, and although the mathematical operations employed in the compilation of results were in and of themselves relatively simple, the argument was already being made before the Civil War that superintendents and their assistants in the central office should have long experience with statistics and their manipulation. The appointment of Joseph C. G. Kennedy, a well-off Ohio farmer who dabbled in newspaper work but distinguished himself mainly as an ardent Whig supporter, to superintend the 1850 Census was in effect the last pure patronage appointment to that post.[46] Kennedy was reappointed to head the following census in 1860, but by that time he had demonstrated a genuine interest in statistics (and had of course gained valuable experience). From then on, superintendents would all have at least some statistical experience behind them when they took office.

An alternative to conventional political appointments had in fact been available for some time, in the form of a tiny community of self-taught but accomplished statisticians. The American Statistical Association was formed in Boston in 1839, and immediately made itself heard on matters relating to

[44] White, *The Republican era*, 10.
[45] C. S. Aron, *Ladies and gentlemen of the civil service: Middle-class workers in Victorian America* (New York, 1987), 108.   [46] Anderson, *The American census*, 35–51.

the census. Its most distinguished early member and president, Dr. Edward Jarvis, penned one of the most systematic and influential critiques of the 1840 Census results (he was echoed by J. D. B. DeBow, editor of *DeBow's Review*, and George Tucker, a political economist at the University of Virginia and author of an influential early statistical history of the US). Jarvis and ASA members Lemuel Shattuck, Nahum Capen and Jesse Chickering had a hand in writing the Census legislation for the 1850 count. Shattuck (himself a respected census taker), was summoned to Washington to place his expertise at the disposal of Kennedy for the 1850 Census and of Kennedy's successor in 1853, J. D. B. DeBow. DeBow engaged Jarvis in 1854 to help compile and publish some of the final volumes.

Although the Boston ASA members remained outside the Federal structure during this period, they enjoyed a high level of influence, and improved the census considerably through their consulting activities.[47] This same group was responsible for publishing the most learned antebellum commentaries on census results, mainly in *DeBow's Review* and *Hunt's Merchants' Magazine*.[48] This influence carried into the post-Civil War period. Jarvis remained involved in census work into the Gilded Age, advising Francis Walker on the Ninth (1870) Census. Walker and his contemporary Carroll D. Wright, who would eventually take charge of the newly formed Bureau of Labor Statistics in 1885, both served terms as president of the ASA. They were among the small pool of public servants with both the integrity and the statistical competence to manage such undertakings (the economist David A. Wells, who originally brought Walker in to head the Bureau of Statistics at the Treasury Department in 1871, was another prominent exemplar who deserves mention).[49]

The existence and availability of such figures did not in any way ensure that their talents would in fact be put to use in the federal agencies. However, again, they represented governmental thinking, such as it was at the time, and their influence did gradually expand. The pattern with such "scientist-officials" seems to have been that once in office (by whatever combination of luck and political maneuvering on the part of their sponsors), their competence and integrity made it difficult for future presidents and members of Congress to replace them with purely political appointments. Walker is a case in point: his reputation and accomplishments probably removed the superintendency of the census from the catalogue of purely political plums earlier than it might otherwise have been removed. It remained a patronage appointment, and a position from which patronage was dispensed, but one which from then on

[47] Davis, "The beginnings of American social research," 163–165; Anderson, *The American Census*, 36–37, 51–52.   [48] Davis, "The beginnings of American social research," 158.

[49] D. Ross, *The origins of American social science* (New York, 1991), 62; "The development of the social sciences," in A. Oleson and J. Voss (eds.), *The organization of knowledge in modern America, 1860–1920* (Baltimore, 1979), 118.

had to be filled by someone with real administrative and statistical experience. Walker's successor as superintendent of the Eleventh (1890) Census, Robert Porter, was appointed partly on political grounds, but he had gained substantial experience under Walker in the central office ten years previously.

Although it represented a drastic change from the status quo, the Walker–Wright–Wells type was itself merely a transitional figure. The very fact that Walker and Wright were given so much personal credit is revealing. Their "impartiality" and "rationality" were understood as a sign of personal probity, not transpersonal institutional necessity. Very shortly after the turn of the century, such "men of affairs," who represented morality in administration, would be eclipsed by the figure of the bureaucratic technician whose personality was irrelevant. In short, during the Gilded Age, the character of governmental "authorities of delimitation" was in the midst of a fundamental and contentious transformation. The task of defining the statistical objects of national governmental discourse rapidly grew more demanding, as did the range of competencies required of those who would carry it out. In the position of the superintendent, the political agendas of Congress were beginning to be shouldered aside by the demands of modern "disinterested" administration.

Grids of specification

The third aspect of the formation of objects Foucault terms "grids of specification," and once again it is helpful to revisit his illustrative examples. Grids of specification were, in the case of early psychiatry, "the systems according to which the different 'kinds of madness' are divided, contrasted, related." Three of the grids he identifies will serve to orient the analysis here: "the soul, as a group of hierarchized, related and more or less impenetrable faculties; the body, as a three-dimensional volume of organs linked together by networks of dependence and communication; the life and history of individuals, as a linear succession of phases, a tangle of traces, a group of potential reactivations, cyclical repetitions."[50] In other words, grids of specification are coordinate systems on which to locate the objects of a discursive formation and their features. If the objects of governmentality are population, activities and resources in their relation to each other and to national territory, and if they are represented statistically in a series of censuses, the relevant grids of specification will be spatial, temporal and social systems of orderly differentiation. The particular statistics reported in the census can be thought of as analogous to particular symptoms of madness, but these symptoms appeared against the background of more fundamental ordering systems. Obviously the most basic structuring coordinate system in a census is spatial: the multilevel nesting of

---

[50] Foucault, *Archaeology*, 42.

state, district and local divisions of the land surface according to which the enumerations are organized. In Chapters 5 and 6, it will be seen that this built-in primacy can encourage the assumption that space is an important independent variable capable not merely of indexing but also of explaining salient features of social order. Here I am merely concerned with cataloguing the different coordinate systems against which the objects of governmentality were specified. Again, since Walker's primary focus was population, I will limit the discussion largely to the grids of specification structuring the population schedules.

Margo Anderson points out that (next to geography) race is the most fundamental grid built into the population schedules of the US census. The constitution already required the census to differentiate between slave (black), free (white) and Indian populations in each state for the purposes of Congressional apportionment.[51] As new categories (for example, insanity) were added to the population schedules with successive enumerations, they were always broken down according to black and white because of the assumed importance of race.[52] Neither emancipation nor the gradual adoption of pro-assimilation policies toward the Indians would reduce the importance of race in the census – "later differentiation of elements of the population would be conceived with reference to these initial distinctions."[53] And the distinctions themselves would multiply. Beginning in 1850, the "color" of a person (other than Indians) could be "white," "black" or "mulatto." For the 1870 and 1880 Censuses, it could be any of those three, as well as "Chinese" or "Indian"; the 1890 Census added "quadroon," "octoroon" and "Japanese" to the range of possible "colors."[54] In the Gilded Age, the Census of Population would be structured so as to highlight issues of what we would call today "ethnicity" and "nationality," which were then thought of as aspects of "race," as though these comprised some of the most important determinants of social behavior and of prospects for a stable social order. As I show in Chapter 7, this "racializing" grid played an important part in constructing the importance of race and ethnicity it took for granted, and in cementing the significance of racial and ethnic definitions of "Americanness" in the twentieth century.

The standard male–female sex distinction was also enshrined in the structure of census schedules as a basic grid of specification.[55] But it probably played a less direct role in perpetuating Victorian gender roles than the racial grid did in perpetuating racial thinking (I will return to this issue shortly). Two other grids would have to be brought into play along with sex in order to uphold patriarchal ideology. Age was one of these. Age and gender together allowed division of the population into distinct groups (adult men, and

[51] Anderson, *The American Census*, 12–13.   [52] Cohen, *A calculating people*, 212.
[53] Anderson, *The American Census*, 12.   [54] Wright and Hunt, *History and growth*, 92.
[55] Anderson, *The American Census*, 36.

women of child-bearing age) taken to represent the productive and reproductive capacities of the American population. A temporal grid also plays an important role in structuring the relation between the census and gender ideology; this grid might be labelled "national progress." The most basic statistic taken to measure national progress was aggregate population growth rate (this was in keeping with the inherited habit, from an earlier, mercantilist colonial era, of considering population a form of national wealth).[56] Population numbers at successive census enumerations, in combination with figures on women of child-bearing age, yielded fertility rates. It was here, in the connection between fertility, population growth and national pride, that the sex, age and temporal grids of specification contributed most to the perpetuation of standard gender roles: women could only play their part in national greatness by remaining at home to bear and raise children. Other components of the temporal grid of specification were statistics of wealth and of agricultural expansion; the pace of enrichment and the westward march were also watched carefully.

Economic class was obscured by the structure of the population schedules, and thus did not serve as a grid of specification. As in other countries, class has been transformed in the US census from an issue of control over the circumstances of productive life to a question of position along a continuum of personal income and location in a neutral table of "occupations." On any of these basic grids, the reality of the "social body" or "nation" would be attested by the presence of regular statistical patterns of one sort or another (recall from Chapter 1 that discovery of such patterns was the original impetus behind the emergence of modern liberal governmental thinking). The indisputably independent, "objective" character of the patterns emerging on these basic grids would endow the grids themselves with an aura of naturalness and objectivity.

The number of "symptoms" or particular statistics structured by these basic grids in the population schedules grew only modestly during the late nineteenth century. While the total size of the census virtually exploded (from five schedules asking a total of 138 questions in 1850 to 233 schedules asking 13,161 questions in 1890), the number of inquiries on the population schedule expanded from 22 to 45. The bulk of the questions added during this period had to do with foreign parentage, citizenship status, more precise specification of family and non-family relationships, and physical and mental disabilities.[57]

This last area of schedule growth suggests one other grid, which I would characterize as "implicit" or "virtual," but which is nevertheless important because it oriented so much of the interpretation placed on census statistics during the late nineteenth century. It might be called the grid of dependency.

[56] Cohen, *A calculating people*, 52–53, 77.   [57] Wright and Hunt, *History and growth*, 87, 92.

The dependency grid is manifested in the categories which imply an inability to function as a productive adult. Dependency was understood by early governmental thinkers as a force acting against national economic productivity. Thus, the relation between the size of the dependent population and the number of working-age men would attract steady comment. Also, in combination with the spatial grid and the grids of race and foreign parentage, this implicit grid would play a powerful role in shaping the American cultural and political response to immigration (see Chapters 6 and 7).

Using the archaeological method as a guide, it has been possible to describe in considerable detail the formation of governmental objects through the medium of the late nineteenth-century US census. Asking about surfaces of emergence has made it possible to see the limitations under which the objects of governmentality were made available to interested parties. Asking about authorities of delimitation has brought to light the importance of larger cultural and political struggles over integrity and competence among public officials as a key context in which to understand how superintendents could construct governmental objects more or less "objectively." Finally, the question of grids of specification has revealed the spatial, social and temporal coordinates making up the "anatomical" structures through which observers understood governmental objects. Together, they tell us in great detail just what characteristics governmental objects could have during this period.

All three aspects of object formation have raised geographical issues which will prove important as the analysis takes shape in the chapters to follow. The problem of time–space convergence or compression, glimpsed in the above discussion of surfaces of emergence, will come more strongly to the fore in Chapter 5, where I explore the spatial politics of knowledge construction. The association of national governmental programs with northeastern elites, which emerged in the discussion of authorities of delimitation, will reappear in Chapters 3 and 7 as one axis along which the contradictions of American national governmentality were (and continue to be) manifested. In particular, it will help explain how even the most impartial exercise of state power could be experienced by average citizens as "despotic."

3

Francis A. Walker and the formation of American governmental subjectivity

---

In this chapter the archaeological method will continue to serve as a guide to the description of the early American governmental discursive formation. If we take Chapter 2 as a detailed outline of the way governmental objects were formed, we may now move on to address the formation of the subjects and subject positions from which the objects could be authoritatively discussed and interpreted during the late nineteenth century. Foucault places such questions under the heading of "formation of enunciative modalities." Although some of the issues (and many of the historical individuals) involved in the formation of enunciative modalities will be familiar from the foregoing account of "authorities of delimitation," the discursive functions of delimitation and interpretation differ in significant ways. They correspond, roughly, to the moments of "observation" and "judgment" in the cycle of social control which I presented in Chapter 1 as the core logic of modern "power/knowledge." For analytical purposes, I treat observation as a prelude to judgment, which in turn forms the basis for regulative decisions. Particularly at the scale of nation-states, delimiting objects is of course always already an interpretive exercise. But the explicitly interpretive discourse in which judgments are rendered occupies a much broader and more diffuse cultural "field" than the delimitative. In this field, struggles between contending interpretations are not welcomed (since the goal is still "objectivity"), but are nevertheless understood to be a more normal, acceptable feature of the discourse rather than automatically a sign of scandal. It was in this broader field of "public affairs" that the "subjects of governmentality" mediated between the objects presented by the census and the ensemble of issues that could be said to structure Gilded Age American middle-class culture. When we have a finer sense of who gained interpretive authority in what contexts, we will have a fairly complete formal "map" of the governmental discursive formation.

This chapter has two main parts. In the first part, I make use of Foucault's archaeological framework to lay out the formation of enunciative modalities in early American governmental discourse. In the second part, I introduce Francis A. Walker more formally and in greater depth, and make the argument that he is ideally positioned to represent the "governmental subject." I close the chapter by suggesting that Walker's career as a governmental subject put a severe strain on the model of "manhood" according to which he tried to live his life. Since his experience in this respect was in many ways symptomatic of a more general transformation of American manhood underway at the time, since (as a result) his governmental program revolved at bottom around his understanding of this issue, and since I use this program as my primary lens through which to view governmentality, it will be necessary to follow this chapter (and end Part I) with an extended excursus on Walker's place in the history of American manhood (Chapter 4).

The public "man of affairs"

As with the formation of objects, so with the formation of enunciative modalities, Foucault identifies three "sub-questions" which help to specify discursive formations in an extremely sensitive manner. First, which individuals are taken as authorized to interpret objects, and what are the sources of the status that gives them authority? Second, from which cultural and institutional sites or positions do these individuals speak? Finally, what sort of epistemological, practical or moral relations do interpretive authorities have with the objects they interpret? Unlike in Chapter 2, here I will not simply take up these questions in order, for two reasons: to begin with, the three questions under consideration here are much more closely intertwined than was the case in the formation of objects. It is hardly possible to discuss the "who" question without being drawn into a treatment of the sites and locations, or vice versa, or indeed to discuss either without immediately raising issues of how interpretive authorities relate to their objects.

In addition, the answers to all three questions changed in fundamental, qualitative ways during the Gilded Age. To be sure, the "surfaces of emergence," "authorities of delimitation" and "grids of specification" described in Chapter 2 changed through the period as well. But the first of these did not change significantly in qualitative terms, and neither did the third (though grids of specification were greatly elaborated in quantitative terms as the number of inquiries on census schedules skyrocketed between 1870 and 1880). Only authorities of delimitation were in the midst of deep qualitative change between the 1870s and the 1890s. By contrast, all three moments in the formation of enunciative modalities, tightly bundled together, underwent basic alterations during the same period. These alterations were complex and mutually implicated in a great many ways, but can nevertheless usefully be described as an extended transition between two endpoints, each personified by a different type of interpretive authority or governmental subject.

The men of affairs who took it upon themselves after the Civil War to interpret census and other factual information for the public and relate it to policy issues were of the same ilk (often in fact were the same people) as the civil service reformers we met briefly in Chapter 2. Their authority to speak on national affairs was to a significant degree self-constructed, out of materials that would come to seem inadequate within two or three decades. Some of the most famous "best men," particularly those in New England, could trade on their status as members of the old established families (Lowells, Eliots and of course Adamses) who had enjoyed public influence since the revolutionary era, and this source of presumed authority continued to echo faintly into the twentieth century.[1] But many men of affairs (for example, German-born Carl Schurz, Kentucky-born Nathaniel Shaler and the Irishman E. L. Godkin) had no such recourse to pedigree. They, like the merchants and newly wealthy industrialists who entered the public forum, drew instead on a combination of real-world experience and the possession of college degrees.

It was not until the early 1870s that the United States as a whole could consistently produce 10,000 college graduates annually.[2] Thus a BA did in fact place one in a small educated elite. Yet advanced degrees were virtually impossible to obtain in the US before the founding of Johns Hopkins University in 1876, and the nature of a BA prior to Harvard's introduction of the elective system in 1869 was not optimally suited to produce competent governmental authorities.[3] Typically the course of study for the BA featured literature, classics and Latin or Greek language, a modern European language, theology, history, art history, mathematics, perhaps some basic astronomy and physics, and in some cases a course of political economy. As most colleges before the late nineteenth century had strong religious affiliations, this curriculum typically culminated in courses on mental and moral philosophy, "where aspects of social life were studied as subdivisions of man's moral behavior in a natural world governed by God."[4] Most of the principles and usages specific to the professions entered by the "best men" would be learned on the job in the manner of apprenticeships. Those who entered the arena of public debate did so as an avocation, usually after having become curious about phenomena or problems encountered in their professional dealings. Doctors like Edward Jarvis often developed an interest in national public health through their work on contagious diseases and sanitation at the urban scale; government officials sought ways to clarify or corroborate the new patterns they thought they saw emerging in the economy and society they were attempting to understand. Some of these men spent enough time and energy delving into the available statistics and literatures to feel that they had acquired the competence necessary to advise the nation.

[1] B. Solomon, *Ancestors and immigrants* (Cambridge, MA, 1956).
[2] US Department of Commerce, Bureau of the Census, *Historical statistics of the United States*, 2 vols. (Washington, DC, 1975), vol. I, 386.
[3] E. Shils, "The order of learning in the United States: The ascendancy of the university," in A. Oleson and J. Voss (eds.), *The organization of knowledge in modern America, 1860–1920* (Baltimore, MD, 1979), 28.
[4] D. Ross, "The development of the social sciences," in Oleson and Voss, *Organization of knowledge*, 109; D. Ross, *The origins of American social science* (New York, 1991), 36.

The formal education of these literary men did not explicitly prepare them for the statistical analyses they would sometimes undertake, but this was not necessarily disastrous. Whatever its failings as a preparation for modern life, this education did at least accustom its recipients to the consideration of national and international issues and long time-frames, especially through the study of history. History, languages and literature also familiarized students with various alternative ways in which states and cultures could approach social problems. Finally, while lacking many of the techniques subsequently developed by the discipline of statistics, the mathematics learned in college at least provided basic algebraic competence.

With these modest but not entirely irrelevant tools, the best men addressed the objects presented in the census results and other national facts primarily as quasi-Archimedean generalists, with a view to synthetic judgment. However, their interpretations were made on the basis of deductive principles as well as inductive reasoning. Statistical facts did not quickly usurp all the authority of *a priori* or deduced principles as numbers assumed increasing importance. Americans were already aware that numbers could mislead (see Chapter 2). And, in any case, without principles and hypotheses it would have been (and remains) impossible to distinguish between more and less significant statistics or statistical relationships in the first place. Inductive reasoning did gain ever stronger prestige during the late nineteenth century and early twentieth century, but only gradually.[5]

Another salient feature of the relationship the best men established with governmental objects was impartiality. Impartiality was a matter of personal reputation and a product of moral conduct. Lacking any precisely formulated procedures such as the "hypothetico-deductive" method which has since become so central to social science, the men of affairs had to demonstrate their impartiality in other time-honored ways. One of these was to assume an attitude of disinterested detachment, to consider matters, as it were, from a distance, to examine all sides of a question. Making meticulous reference to established authorities also reflected well on one's impartiality, as did reproducing whatever reported facts and figures served as the basis for one's argument. These appurtenances of "sound scholarship" were of course not resorted to purely out of a cynical desire to have one's opinion prevail; they expressed also a genuine desire to establish truth. But the thorough entanglement of personal reputation in the outcome of debates perhaps lent an extra urgency to such strategies. The most extreme act of impartiality, and at the same time the riskiest, was the frank criticism or abandonment of one's own earlier position on some subject. While such acts powerfully suggested that their perpetrators were committed to the truth above all else, they could also suggest a certain inconstancy or changeability which might endanger the perception of moral strength. Perhaps for this reason, fundamental self-criticism

5 Ross, *The origins of American social science*, 428–430.

was a step seldom taken by late nineteenth-century American men of affairs. In their world, the advantages of such epistemological self-effacement were too dearly bought. A disinterested approach to the interpretation of governmental objects was still most fundamentally an expression of distinctive personal qualities, and it would not do to efface the person whose reputation rested on those qualities.

It is not a coincidence that the early mass-market magazines in which the most widely read diagnoses of the state of the nation appeared had, by 1870 or so, adopted the practice of publishing contributors' names. Before 1850, a much larger proportion of social, political and economic commentaries were written by editors. According to Frank L. Mott, there were in the antebellum period "few practiced writers in America,"[6] and many of those few were "gentlemen authors" who viewed writing as a leisure activity, wanting neither payment nor notoriety. By the late 1880s, the serious journals staked much on the reputations of their authors, and the authors, in turn, staked far more of their reputations on the articles they wrote.

The transitional period, part I: clubs and magazines

To see how the men of affairs "worked themselves out of job" as governmental interpreters, we must turn from the first and third to the second aspect of the formation of governmental subjectivity: from the people and their relations with the objects to the institutions from and through which they presented their views. It was largely through the efforts of the men of affairs that the literary clubs and monthly or quarterly magazines which carried much of the early governmental discourse surrendered pride of place by the 1890s to research universities, professional associations and specialized social science journals.

Whatever their particular geographical origins, members of the northeastern educated elite plied their professional trades and launched their public careers where there was a market for their services and ideas: predominantly in the established cities of the eastern seaboard. Boston, Philadelphia and New York City formed the core urban network of publishing and correspondence for this infant community of "national" governmental thinkers. But there were important links also with Washington, DC (of course), with cities somewhat lower in the urban hierarchy but still important (Baltimore, Richmond, Charleston, SC, Cincinnati, Lexington, KY), and with college towns of greater or lesser importance (New Haven, Princeton, Providence, Ithaca and Hanover).

---

[6] F. L. Mott, *A history of American magazines*, 5 vols. (New York [vol. I], 1930; Cambridge, MA, 1938 [vols. II, III]; 1957, [vols. IV, V]), vol. II, 25. This remains the definitive, and by far the most exhaustive work on this feature of American history. I rely on Mott (unless otherwise indicated) for the factual details in the narrative to follow, but the arrangement of these details into a narrative is my own doing.

The most important institutions through which Civil War era men of affairs organized early governmental discussions were literary reading clubs and monthly or quarterly magazines. In the first half of the nineteenth century, one of the great difficulties involved in publishing a magazine was finding a sufficiently large number of contributors to provide a steady stream of articles. Quite early on, serious American magazines were thus often associated with compact literary clubs whose members were expected to produce publishable material regularly. However, even the clubs often failed to generate enough material, with the result that editors frequently did much of the writing. But the club structure did provide a core organized market for serious writing. By 1865, magazines routinely offered reduced subscription rates to "home clubs" as well as elite literary societies. In this arrangement, anyone who could organize five, ten or more subscribers would receive a free subscription, and the club members would enjoy a significant discount. Before 1825, the contributor and market networks formed by these literary and home clubs were almost entirely local, rarely extending more than 50 miles from the publishing house. This changed in generally predictable ways as transportation was revolutionized by the railroad, but even in the late 1880s, only New York, Boston and Philadelphia were home to genuinely national magazines.

As the preeminent national forum, the magazine had its heyday roughly from the end of the Civil War to the turn of the century. During this period, mass circulation magazines "came to reflect more directly than ever before the current thought of the country not only in literature and the arts but in politics, economics and sociology."[7] William F. Poole, compiler of *Poole's Index of Periodical Literature*, claimed in 1882 that "periodical literature was never so rich as during the past thirty years. The best writers and the great statesmen of the world, where they formerly wrote a book or a pamphlet, now contribute an article to a leading review or magazine, and it is read before the month is out in every country in Europe, in North America, in India, Australia, and New Zealand."[8]

By the 1870s, Boston and New York City dominated serious magazine publishing, the former steadily losing ground to the latter in terms of number and circulation of magazines but nevertheless retaining disproportionate prestige.[9] The key magazines in which one could find weighty discussions of governmental issues were *Atlantic Monthly, Harper's, Scribner's* (later *Century*), *Putnam's*, the *North American Review, International Review, Forum* and *The Nation*. One could also include *Princeton Review, Lippincott's, Arena* and

---

[7] Mott, *A history of American magazines*, vol. III, 280.
[8] W.F. Poole, *Poole's index to periodical literature*, 3rd ed. (Boston, 1882), iv.
[9] E. Sedgwick, "The Atlantic Monthly," in E. Chielens (ed.), *American literary magazines* (Westport, CT, 1986), 53; Mott, *A history of American magazines*, vol. III, 25–29. The generalizations I draw in the passages to follow are based on material gleaned from Mott, and from the appropriate entries in Chielens and A. Nourie and B. Nourie, (eds.), *American mass market magazines* (Westport, CT, 1990).

Galaxy. The first four of these were general literary magazines in which serious non-fiction played only a secondary role. The Boston-based *Atlantic* (est. 1857) was the most serious of the four. It provided the premier outlet for the Olympians of New England high culture. The other three were New York based publishers' magazines, each of which sandwiched serious writing between some combination of literature, light news, travel, humor, advertising and attractive illustrations. The independent *Nation* (est. 1865) also included literature but was more strongly committed to social, economic and political commentary. The *North American Review*, established in Boston in 1815 was the oldest, most august and serious of the lot, offering lengthy treatises on public affairs. Together with *International Review* (est. 1874), it was the chief American conduit for news and analyses from Europe and other parts of the world. *Forum* (and its lesser cousin *Arena*) were devoted to current affairs, particularly discussions of reform movements. Together these magazines constituted the main vehicle for the thought of the best men. For example, a cursory scanning of *Poole's Index* for the 1880s and early 1890s reveals a concentration of articles about the census by the country's most knowledgeable commentators in the *Nation*, the *North American Review* and *Forum* (as well as a special series in *Popular Science Monthly*). Other topics of current interest treated with some regularity in these and other less central journals included Indian affairs, reconstruction, Darwinism, immigration, tariffs and free trade, labor unionism and socialism, civil service reform and partisan politics more generally.

The size of the audiences reached through these magazines varied widely, with *Harper's* and *Scribner's* showing circulations in the hundreds of thousands while the *Atlantic*, the *Nation* and the *North American Review* all struggled at different times during the period to surpass or stay above 10,000. Circulation of the more popular publishers' magazines began to include significant newsstand sales during the Gilded Age, but the more highbrow publications continued to rely almost exclusively on subscriptions. The size of the audience most acutely interested in governmental matters is probably well estimated with a glance at the circulation history of the *North American Review* through the 1880s. The *Nation* would also serve as a plausible guide, having quickly achieved notice for the prestige of its authors and the quality of its writing, but the *NAR* was considered "the highest, most impartial platform upon which current public issues can be discussed."[10] It was "read by the leading men and was available at all the important reading rooms. Over a hundred copies went to England" even before the Civil War had ended.[11] Its circulation climbed from 7,500 to 17,000 between 1880 and 1889. If we allow for reading room and library subscriptions, it seems reasonable to place the

---

[10] D. Straubel, *"North American Review,"* in Nourie and Nourie, *American mass market magazines*, 334.   [11] Mott, *A history of American magazines*, vol. II, 232.

size of the audience between 20,000 and 50,000. Of these, some unknown proportion would have been less keenly interested in American governmental affairs than in other topics covered, and would perhaps have passed over discussions of the census when they appeared. In any event, it is probably safe to estimate the effective public for such writing as somewhere in the lower tens of thousands nationwide.

The geographical distribution of this public and of its publications played an important role in limiting the possibility for an effective national discourse of governmentality. Interpretations of the national body politic flowed from Boston and New York City down the urban hierarchy, as well as southward and westward, but very little serious discourse made the return trip. For obvious reasons (especially when the topic was reconstruction), southern writers and editors understood the major national magazines as organs of northeastern propaganda, not as the impartial vehicles for definitions of the "general interest" which they purported to be. Similarly, many westerners strongly resented the image of western lawlessness and immorality they felt the eastern press had unfairly constructed, and were often enraged when these distant arbiters of sound opinion presumed to dictate policy on such inherently western matters as Indian affairs and federal land management.[12] Although the specifics of the geography of governmental discourse would change in significant respects, its association in general with contestable eastern interests would stick into and through the twentieth century. Thus programs devoted to the general improvement of the population and economy on a nationwide basis would continue to be understood as unwelcome impositions from distant ruling centers.

There was another problem with the public arena the men of affairs had built for themselves in contributing to national magazines. Simply put, the problems facing the nation, and the volume of information and commentary on these problems, was beginning to render the mass market magazine format inadequate. One minor but telling symptom of the strain was the appearance of the magazine *Review of reviews* in 1890. Its founder, W. T. Snead, explained its inauguration in this way: "of the making of magazines there is no end. There are already more periodicals than anyone can find time to read. That is why I have added another to the list. For the new-comer is not a rival, but rather an index and a guide to all those already in existence."[13] William F. Poole began publishing his mammoth index to American (and some British) periodical literature in 1882 for similar reasons. It was more generally an era of attempts to gain control of exploding literatures of all sorts, in books as well as magazines. The Dewey decimal system was worked out in the 1870s; the medical literature was first indexed in 1880; law in 1897. By the mid-1890s,

---

[12] *Ibid.*, vol. III, 59.
[13] Quoted in V. Schwartz, "*Review of Reviews*," in Nourie and Nourie, *American mass market magazines*, 435.

a comprehensive index to federal government documents had been prepared; *Books in print* first appeared in 1900.[14]

Like reading, writing about public affairs had become more difficult. Unlike in the ante-bellum period, it was now much more difficult for a single author to present a comprehensive and convincing treatment of a major national issue in a single magazine article. The 1880s were marked by ever more frequent recourse to symposia, in which a range of prominent thinkers would present arguments on a chosen topic. The *North American Review* took the lead in this practice during the 1880s, and two new journals (*Forum*, est. 1886 and *Arena*, est. 1889) made symposia trademark features.[15]

This may seem an unremarkable development, but seen in the context of changes in the formation of enunciative modalities, it acquires more importance. I contend that the rise of symposia during the 1880s was in a sense the final heroic attempt by the best men to maintain their interpretive grasp as traditional public men of affairs of the features of the American socioeconomic order. As the issues grew ever more complicated and the fund of available information ever more voluminous, it became necessary to improve on the outdated model of the single learned interpreter. Since the alternative to the man of affairs was still only imperfectly discernable before the 1890s, it made eminent sense at the time to respond to a more complex world by mobilizing larger numbers of the best men to comment on it. But it would soon become evident that bringing in more generalists was not the most effective solution. Indeed some of the symposiasts by the 1890s were no longer easily identifiable as "men of affairs," but personified instead a new approach to public issues.

The logic of convening symposia in the magazines had its parallel in the state in the form of special commissions. The practice of appointing special commissions of distinguished gentlemen to investigate problems or monitor events preceded the Gilded Age and would survive into and through the twentieth century. But I would argue that it had a special significance during the late nineteenth century, in so far as the federal government retained many of the features of a "state of courts and parties."[16] As Michael Mann points out, the United States had a relatively puny and underdeveloped administrative apparatus given the size of its economy and population.[17] The socioeconomic processes which posed such an epistemological challenge to mid-century social thinkers constituted an equally serious administrative challenge to these feeble structures. Lacking a developed set of procedural regulations, and

---

[14] J. Higham, "The matrix of specialization," in Oleson and Voss, *Organization of knowledge*, 14–15; L. Schmeckebier, *The Government Printing Office: Its history, activities and organization* (Baltimore, MD, 1925), 15.

[15] Mott, *A history of American magazines*, vol. III, 282; vol. IV, 51, 207–216, 354–356.

[16] S. Skowronek, *Building a new American state: The expansion of national administrative capacities, 1877–1920* (New York, 1982).

[17] M. Mann, *The sources of social power*, vol. II, *The rise of classes and nation-states, 1760–1914* (New York, 1993), 366, 369.

lacking public support for any but a handful of permanent bureaucracies, the state had to find other ways of handling the problems forced upon it in a manner that would be as impartial as possible. Thus, as with social commentary, the most obvious solution given the existing resources and cultural assumptions was once again a group of the best men. A classic example from the early Gilded Age was the Board of Indian Commissioners assigned to bring Christian rectitude to bear on the activities of the corruption-plagued Bureau of Indian Affairs. Later in the period, religion would recede into the background as a source of rectitude, and such bodies as the Civil Service Commission would attempt to build their reputations for fairness out of the secular impartiality of their members. But as with symposia in the magazines, special commissions were increasingly hampered in their usefulness by their lack of expertise. Their existence alongside the early signs of governmentalization of the American state signals a continuing respect for "reputation," but reputation was no longer enough.

The transition period, part II: associations, universities and specialized journals

As the treatment of public issues in the magazines was evolving in the manner just described, another set of institutions, organized by the men of affairs but carrying the seeds of their demise, was poised to transform the enunciative modalities of governmental discourse. The first of these to appear was the professional social science association. The American Statistical Association (discussed in Chapter 2) was in a broad sense the model for later associations. Although there were no professional statisticians in the modern sense until some fifty years after its 1839 founding, the ASA was animated by a seriousness of purpose and a national orientation that would mark subsequent associations. Even at the time of its founding, it strove not to be "preoccupied by local affairs or wasting time in mutual self-admiration over Massachusetts men and things."[18] This showed an acute awareness of the pitfalls of broadcasting supposedly national truths from the Boston area. The 1839 by-laws insist that the operations of the ASA be "as general and as extensive as possible and not confined to any particular part of the country."[19]

The most important organization to appear after the Civil War was the American Social Science Association, founded in 1865. The ASSA brought together many of the most prominent best men from their scattered posts in fledgling government bureaucracies, in college teaching and in the professions. Its leadership was composed of important members of the New England elite, in particular natural scientists (Louis Agassiz, B. F. Pierce) urging higher

---

[18] J. Koren, "The American Statistical Association," in J. Koren (ed.), *The history of statistics: Their development and progress in many countries* (New York, 1918), 4.   [19] *Ibid.*, 3–4.

standards of scholarship in social research, and influential university reformers such as Charles W. Eliot of Harvard and Daniel Coit Gilman, first president of Johns Hopkins University.[20] According to Dorothy Ross, the program of general edification first pursued by the ASSA was based on the same determination to preserve the leadership role of the men of affairs that characterized the discourse in the magazines. As in the mass market journals, the members of the ASSA at first promoted specific reforms (particularly Civil Service Reform) and were thereby clearly associated with the "mugwump" faction of the Republican Party. Thus for a time their claim to represent the "national interest" was difficult to substantiate. Within twenty years of its founding, however, fundamental changes would consign the ASSA's "genteel" approach to the dustbin.[21]

The structure of the ASSA would provide the setting for its demise. It was divided into four sections: education, public health, social economy and jurisprudence. As members of these sections met periodically to discuss important issues and digest the latest information provided by censuses, urban surveys and the like, they began to define central issues peculiar to the study of ever-more-specific aspects of American society. This process of incipient specialization was given an accelerating kick in the 1880s as young Americans who had done graduate work in Germany returned to push for more formal and specific disciplinary definitions.[22] The result was a series of splits which gave birth to many of the modern organizations still in existence today.[23] The American Economic Association hived off from the ASSA in 1885. Along with the independently founded American Historical Association (1884), the AEA formed the matrix for the birth of the national sociology and political science associations. By 1905, history, economics, political science, sociology, psychology and anthropology all had national associations.[24] These societies represented a more plausible "effort to break out of the boundaries set by locality and to reach across space into national communities," in part because membership was increasingly dependent only on interest and training, not on acquaintance with elites from a particular regional, social or economic background.[25]

Specialized professional journals proliferated during the same period, sometimes predating but often published by national associations: the *Publications of the American Economic Association*, the *Quarterly Journal of Economics* and the *Political Science Quarterly* all appeared in 1886, the *Publications* [later *Journal*] of the American Statistical Association in 1888, the *Annals of the American Academy of Political and Social Science* in 1890, the *Journal of Political Economy* in 1892 and the *American Journal of*

[20] Ross, "Development of the social sciences," 109–110.
[21] A. Oleson and J. Voss, "Introduction," in Oleson and Voss, *Organization of knowledge*, xiii.
[22] Shils, "The order of learning," 34.
[23] Ross, "Development of the social sciences," 110, 111, 113, 116.   [24] *Ibid.*, 108.
[25] Shils, "The order of learning," 35; Oleson and Voss, "Introduction," vii.

Sociology in 1895.[26] These were not at first the highly specialized outlets they would soon become. Although the various social sciences were growing ever more distinguishable in the central problems they treated, they had as yet not formulated advanced methodologies comprehensible only after years of training (the more mathematically sophisticated marginalist approach had already made its debut in economics, but had not yet established itself at the core of the discipline). The specialized journals and associations were still predominantly located in northeastern cities, but were associated not so much with a particular class or political persuasion as with a new and powerful institutional form: the modern research university.

Like the associations and journals for which they provided the anchoring points, the research universities were a creation of the men of affairs which ended up contributing to their intellectual marginalization. The founding of Johns Hopkins University, as an institution whose chief mission was to produce advanced research and train scientists in graduate programs, was a key event in this respect as well.[27] Clark University in Worcester, Massachusetts (1887) and the University of Chicago (1892) adopted similar missions, and the older established colleges began during this period to offer graduate degrees. Charles Eliot's 1869 decision to institute the modern elective system for Harvard undergraduates was soon copied as well, allowing academic specialization to begin earlier in the careers of American students. By the 1890s, the American social science disciplines had ceased to rely on study in Germany for the training necessary to staff the growing networks of university departments. In 1870, no Master's Degrees and only one Ph.D. were granted by American universities, and the few non-German trained social scientists teaching courses in individual social science disciplines (e.g., Francis Walker, James C. Dunbar and Andrew White) were self-taught men from the ranks of the northeastern elite.[28] By 1885 the numbers of indigenous MAs and Ph.D.s had risen to 1071 and 77, respectively. Ten years later, the figures were 1,334 and 272.[29]

Especially with the rise of the University of Chicago and the spread of specialized graduate training to the land grant universities of the midwest, the basic geography of governmental discourse was altered considerably. The journals and conferences, as well as the horizontal movement of students between schools, effectively integrated a set of national communities. But the center of gravity was still northern and eastern, and the social sciences were already by the beginning of the new century developing a new kind of

---

[26] Mott, *A history of American magazines*, vol. IV, 180–192.

[27] C. Lucas, *American higher education: A history* (New York, 1994), 173; J. Brubacher and W. Rudy, *Higher education in transition: A history of American colleges and universities*, 4th ed. (New Brunswick, NJ, 1996), 178; Shils, "The order of learning," 28.

[28] Ross, "Development of the social sciences," 110.

[29] US Department of Commerce, Bureau of the Census, *Historical statistics of the United States: Colonial times to 1970*, 2 vols. (Washington, DC, 1975), 386.

"distance" from the general populace: the distance conferred by expert training. As the professional journals solidified their status as the outlets for the
most qualified authorities, mass market magazines became a less prestigious
platform for high debate. The journals which still attempted to inform the
public on serious topics in anything more than an occasional and highly
watered down manner (for example the *Nation* or the *North American Review*)
were pushed to the margins of the market and left to struggle with meager circulations.

The expert social scientist

The effect of these structural changes in the enunciative modalities of
American governmental discourse was to create a new species of governmental subject: the expert social scientist. The public man of affairs was not
thereby completely banished from public life, but he was certainly ushered to
the periphery.

The signal characteristics of modern expertise are so well rehearsed as to
need little in-depth discussion here. The expert, as opposed to the man of
affairs, is a specialist with intensive training in a narrowly defined field of
study. Through a vast division of labor, social science specialists build a complete overview of social reality only in the aggregate. Thus the expert relates
to the objects of governmentality more in an analytical than in a synthetic
manner. The implicit ideal of comprehensive individual epistemological
mastery of the national social body gives way to a less heroic program of collective coverage. In principle, the personal identity of a social scientist is
utterly irrelevant to the performance of research; this irrelevance is carefully
maintained through strict adherence to scientific method and its well-defined
decision procedures. The impartiality once flowing from individual moral rectitude is increasingly located in generally agreed upon methods of inquiry.
This is the second respect in which the experts' relation to governmental
objects differs from that of men of affairs. In principle at least, expertise so
constituted fits very nicely into the logic of power/knowledge at the heart of
modern governmentality, for it removes much of the capriciousness from
policy judgments and furthers the perception of rational government as
"automatic" and impersonal (see Chapter 1).

The differences between social science experts and their "literary" predecessors are clear from this fleeting sketch, but it is also possible to see continuities
in the transition between the two. Edward Shils detects a "stern moral over-
tone" in specialization. It brooks "no trifling, no self-indulgence . . . In sum,
specialization was consistent with the secularized Protestant Puritanism of the
quarter century preceding World War I."[30] Something like this perception

---

[30] Shils, "The order of learning," 33.

probably allowed those older elites (such as Francis Walker or Daniel Coit Gilman) who were most active in building the new epistemic infrastructures to remain on familiar moral ground while doing so.

The new expert social scientists were in fact not so abstracted from the realm of cultural, economic or political partisanship as the foregoing idealized description might suggest. One problem they faced was that specialization had long gone "against the grain" of American intellectual culture; it "violated the American ideal of the untrammelled individual and its corollary, the Jack-of-all-trades." The specialist "affronted the egalitarian values: he dealt in secrets only a few could share."[31] This pervasive cultural distrust had abated somewhat by the Progressive Era that began roughly with the new century.[32] But it was probably among the factors which ensured that unlike in other contemporary "modern" countries (for example, France or Germany), the vast majority of trained experts on public policy in the United States would make their institutional homes outside the state structure.[33] On this logic, the public and its Congress were happier to see expertise confined in universities, with only contingent connection to the levers of governmental power. Thus the program favored by Carroll D. Wright, in which advanced social science training would take place within functioning bureaucratic structures and connected training schools (i.e., more on the French pattern), probably never stood much of a chance.[34]

Another result of the lingering distrust of expertise was the frequent and increasing recourse to statistics noted by Theodore Porter (see Chapter 1). The publicity of statistics, their availability for scrutiny by anyone who cared to inquire, could in principle go some way toward dispelling the damaging aura of mystery and secrecy attached to expertise.[35] To the extent that they gained influence in governmental matters, social scientists came under pressure to rely more heavily on statistics, and this dynamic probably contributed to the intensifying "scientism" noted by Dorothy Ross. As scientists came to value a high public opinion of science, they placed increasing emphasis on disciplinary consensus, and tended increasingly to stay away from advocacy of sweeping reforms or acceptance of government posts.[36]

However, the consensus positions they reached were not as reliably pro-capitalist as had been the opinions of the best men who preceded them. The new social scientists were far more often from lower-middle-class (perhaps immigrant) families, often the first in their families to attend university. They had less reflexive allegiance to the socioeconomic status quo, and were more likely

---

[31] Higham, "The matrix of specialization," 4.

[32] R. Wiebe, *The search for order, 1877–1920* (New York, 1967).

[33] B. Silberman, *Cages of reason: The rise of the rational state in France, Japan, the United States, and Great Britain* (Chicago 1993).   [34] Ross, "Development of social science," 118.

[35] T. Porter, *Trust in numbers: The pursuit of objectivity in science and public life* (Princeton, NJ, 1995).   [36] Ross, "Development of the social science," 125–130, 123.

than their elite predecessors to favor state intervention in the economy. For example, the young German-trained economists who formed the AEA in 1885 did so in part to establish a platform from which to attack the *laissez-faire* consensus that still held sway in their profession.[37] Thus the fourth way in which their relation to the objects of governmentality differed from that of the men of affairs was in its sense of involvement rather than detachment. While the best men had advocated no end of reforms and changes, they did so from a distance, and the same mindset that allowed them to take this distance for granted restricted the range of deliberate, state-led social changes they could easily imagine.

The deepest continuity between men of affairs and the experts who succeeded them, however, was that they were *men*. The reins of social science as it was transformed into its modern academic form remained in male hands throughout the transition. The story of how male control of the enunciative modalities under consideration here was maintained sheds crucial light not only on the subject positions occupied by Francis Walker, but also more generally on the theorization of governmentality given in Chapter 1. The account I have given in this chapter describes the enunciative modalities that came to be associated with modern American social science. But the definition of social science as an academic enterprise was a political move by male social scientists to exclude women from its inner sancta. Recent feminist scholarship has drawn the attention of historians to a competing model of social science and regulation championed by a prominent group of women social scientists.

Scholars have long been aware of the key role played in the early Progressive Movement by highly trained women such as Florence Kelley, Sophinisba Breckenridge and Elsie Clews Parsons, among others.[38] The research and advocacy undertaken by these and other women in Chicago and other cities spearheaded the reform movements that eventually gained presidential patronage and national support under Theodore Roosevelt. However, all of this activity has traditionally been understood by scholars as social activism or social work, not social science.[39] Because it was locally focused, strongly oriented toward reform and often undertaken from non-academic institutions such as Jane Addams's Hull House in Chicago, the assumption has been that it was qualitatively different from academic work. However, Florence Kelley and others were highly trained, often holding Ph.D.s from the University of Chicago, Yale or one of the other universities gradually opening their doors to women. And the research they produced was every bit as statistically oriented and painstaking as that produced by their male counterparts on the new social

[37] *Ibid.*, 114–116.
[38] H. Silverberg, "Introduction: Toward a gendered social science history," in H. Silverberg (ed.), *Gender and American social science* (Princeton, NJ, 1998), 3–32; S. Koven and S. Michel (eds.), *Mothers of a new world: Maternalist politics and the origins of welfare states* (New York, 1993).
[39] E. Fitzpatrick, *Endless crusade: Women social scientists and progressive reform* (New York, 1990), xiii.

scientific university faculties. Mary Jo Deegan argues that *Hull House Maps and Papers*, a detailed statistical and cartographic study of immigrant neighborhoods in Chicago researched and published by Kelley and other Hull House women, "mark[ed] the intellectual birth of Chicago sociology."[40] Albion Small and other Chicago school founders supported and made use of data produced in these and other studies by women social scientists as they developed the contours of their social scientific practice.

But these men did not admit their female colleagues into the institutional home they were building for modern social science. The concentration of trained women social scientists outside the academy was the result of their having simply and brazenly been denied access to faculty positions in the newly formed academic social science departments at Chicago, Johns Hopkins, and elsewhere.[41] Thus the definition of social science as an academic enterprise distinct from the more openly "moral" sphere of "social work" is a gendered definition representing a political act of exclusion. The older northeastern elite was struggling to adapt to the new urban-industrial society in a range of ways, most aimed at preserving as many dimensions of the old social hierarchies as possible. As middle- and upper-class native-born white men, they sat atop at least three of the most important hierarchies, and thus attempted to preserve class, race and gender distinctions in whatever ways they could. The discursive formation of national governmentality was gendered not merely in the formation of objects but also in the definition of the subject positions assumed to qualify one to speak on governmental matters.

This suggests that a feminist approach could render the concept of governmentality more analytically sensitive. It is worth taking the time to spell out some of the potential changes. In Foucault's original formulation, and in much of the subsequent research inspired by it, the differently gendered character of different programs of governmentality has been noted on occasion, but hardly explored at all. Foucault himself, at least in his seminal essay, seems to have assumed that the overarching issue of "government" was always government by men, in the family as well as in public life. Although, as in his earlier work on biopower, he was well aware that social regulation had specific effects on women, the agents of social regulation remained exclusively male.[42] Jacques Donzelot, in his intricate study *The policing of families*, notes the association between women and moral authority in late nineteenth-century France, but also focuses on the male theorists of "social economy," giving only derivative agency to women.[43] This may be in part a matter of differences in national context. The generally more centralized institutional structure of French social regulation would have made it far more likely that men held

[40] M. J. Deegan, *Jane Addams and the men of the Chicago School, 1892–1918* (New Brunswick, NJ, 1988), 66.   [41] Fitzpatrick, *Endless crusade*, 9.

[42] M. Foucault, "Governmentality," in G. Burchell, C. Gordon and P. Miller (eds.), *The Foucault effect: Studies in governmentality* (Chicago, 1991), 87–104.

[43] J. Donzelot, *The policing of families* (Baltimore, MD, 1997 [1979]).

the reins. The American feminist movement was certainly in full swing much earlier than its French counterpart. Be that as it may, the American experience casts the basic contours of the logic of governmentality into a new light. In the United States, as Stephanie Coontz notes, the moral authority which the Victorian era domestic ideology granted to women was subsequently used by them to try to infuse higher standards of morality into the public realm of economics and politics.[44] The anti-slavery and temperance movements constructed by women were supplanted or supplemented toward the end of the century by wider women-led campaigns against urban and industrial ills of all sorts (pauperism, child labor, etc.). These campaigns were backed by statistical social science, not merely moral sentiment. Thus it is not surprising that the "social economy" section of the early ASSA was a site of notable influence by women reformers.[45] In other words, the distinction Procacci establishes between "abstract" political economy and "moralistic" social economy (see Chapter 1) was a gendered distinction in the American context.

In the essay introducing their compilation on governmentality, Andrew Barry, Thomas Osborne and Nikolas Rose portray independence and distance from social activism as central to the role played by academic social science. In Chapter 1 above, I borrowed their formulation to explain how normalization is carried out in an ideal-typical modern national system of governmentality. Social science, by virtue of the distance it puts between itself and the specific interests struggling for influence within the context of the state, achieves the substantial credibility that allows its recommendations to be taken seriously. But if this distance already has a clear gender politics behind it, as Kathryn Kish Sklar, Helene Silverberg, Ellen Fitzpatrick and other feminist scholars argue, then the whole logic of governmentality can be seen as masculinist.[46] Certainly the particular instance I explore in this study would reinforce this insight. The upshot of all this is that gender should become a more prominent component not merely in concrete studies of governmentality, but also in the evolving theorization of the concept.

Francis A. Walker and the governmental discursive formation

We now have at our disposal an admittedly selective but serviceable account of the formal characteristics of the American governmental discursive formation during the last decades of the nineteenth century. Chapter 2 laid out the

---

[44] S. Coontz, *The social origins of private life: A history of American families, 1600–1900* (New York, 1988).   [45] K. Kish Sklar, "*Hull House Maps and Papers*: Social science as women's work in the 1890s," in Silverberg, *Gender and American social science*, 129.

[46] M. J. Deegan, *Jane Addams and the men of the Chicago School, 1892–1918* (New York 1988); E. Fitzpatrick, *Endless crusade: Women social scientists and Progressive reform* (New York, 1990); H. Silverberg (ed.), *Gender and American social science* (Princeton, NJ, 1998); K. Kish Sklar, "The historical foundations of women's power in the creation of the American welfare state, 1830–1930." in Koven, S. and Michel, S. (eds.), *Mothers of a new world* (New York 1993), 43–93.

surfaces of emergence, authorities of delimitation and grids of specification through which the US Census inscribed the national social body as an object of analysis and potential regulation. The historical specifics of these three "dimensions" of the national body constituted a set of limitations on how this body could be understood, and on the prospects for plausibly "objective" empirical knowledge of it. The bulk of the present chapter has been devoted to showing how the constitution and positioning of governmental subjectivity, as well as its relations to the objects of governmentality, were transformed through the late nineteenth century. Together, the formation of objects and of enunciative modalities had the effect of tying together more strongly the first two moments in the cycle of governmental social control. While the first moment, "observation" had been carried out since the birth of the republic (at least in the form of the US Census), it had not played a particularly profound role in the second moment, normative judgments about what to do, judgments that would dictate specific policy decisions. When it did play a role, it was usually as ancillary support for interested judgments already made. By the 1890s, as the "authorities of delimitation" achieved a more serviceable reputation for objectivity, and as the interpretation of the national socioeconomic order increasingly came under the auspices of modern social science, technologies of observation such as the census acquired a more central, determinative role in the formation of governmental judgments. To be sure, federal decision-making remained (as it remains today) a complicated mix of the openly "political" and the apparently more "rational." And as Chapter 2 made clear, even the seemingly rational statistical form of the governmental object predisposed judgments toward issues of region and race. Finally, there was clearly also a contentious politics of space shaping the discursive formation. But despite all of this, the basic outlines of a large-scale cycle of social control were slowly taking form: the governmentalization of the state was slowly creaking into gear.

I propose now to personify this whole process for the purpose of keeping my analysis manageable. Although I have already significantly narrowed the question of late nineteenth-century American governmentality by limiting myself to discourses loosely surrounding the census, the present chapter should make it abundantly clear that the discursive formation with which I am concerned remains a vast, messy business. To do comprehensive justice to early governmental discussions of social statistics set in their context would be a far greater task than one could accomplish in a single volume. Thus I recruit a single gentleman as a "symptomatic figure," a representative "governmental subject" through whose eyes and through whose career I can explore a representative governmental program. As I indicated earlier, Francis A. Walker is that gentleman. Merely on the basis of his importance to the Gilded Age censuses, he automatically occupies a central place in the emergence of governmentality. But he is in fact far more thoroughly qualified than that, having occupied at one time or another in his remarkable career every important subject position in the whole discursive formation I have just described. In addition, Walker had a prominent hand in the transformation that took place in the nature of these subjects positions from the immediate post-Civil War period to the 1890s. He inhabited the typical subject positions both of the early and of the late Gilded Age governmental discursive formation. Walker was of course not the only man of affairs to have a prominent and varied public career, but his was most precisely centered on those institutions and issues directly involved in the rise of rudimentary governmentality. Once again, the point of recruiting Walker is not to write his biography. As this chapter should have made clear by now, I am interested in his "structural" characteristics. By attributing to him a single, coherent "governmental program," I am already departing from the conventions of biographical writing. His program was something of which he was probably not aware beyond a general desire to be consistent. In fitting its pieces together, I will obviously have to go well beyond his own understanding of what he was doing and thinking; but my construction should not be mistaken for an impartial historical description of his intellectual life.

This study is not the first to approach Walker as a symptom of his times. In 1972, Richard J. Sidwell completed a doctoral dissertation in economics at the University of Utah which employs a sociology of knowledge perspective to cast Walker as a symptomatic figure. According to Sidwell, Walker's thought flowed from his position as a "class ideologist" representing America's business, political and cultural elite, and more specifically, from his status as a "Brahmin ideologist," representing the concrete community of elites centered in Boston.[47] Walker was certainly both of these, and some of what follows in this chapter amounts to a reconfirmation of Sidwell's basic insight. But in the quarter century since Sidwell completed his thesis, the range of interpretative tools available for linking the ideas of individuals to features of their socio-historical contexts has grown exponentially, both in scope and in subtlety. Thus the present study can be seen as, among other things, an updating and extension of Sidwell's work.

Walker's credentials as one of the "best men" were impeccable.[48] He was born in Boston in 1840 to an old New England family, and grew up in North Brookfield, MA, where they moved when Francis was three. His father, Amasa Walker, was an ex-minister turned businessman who became a respected political economist in his later years. The elder Walker taught political economy at Amherst College, and after Francis graduated from that school in 1860, he briefly took over his father's teaching duties at his alma mater. Political

---

[47] R. Sidwell, "The economic doctrines of Francis Amasa Walker: An interpretation," Ph.D. thesis, University of Utah (1972), 18–24.

[48] Much of the information presented in the following passages concerning Walker's career was gleaned from the only extant biography, James P. Munroe's *A life of Francis Amasa Walker* (New York, 1923).

economy was still a somewhat unusual calling, and his early interest in it had not prevented the younger Walker from acquiring the classic "literary" education. Walker's involvement in the trappings and social circles of New England high culture were lifelong; in later years he spent considerable time with the Holmeses and the Adamses when in Boston, joining them, for example, in the elite congregation of the Reverend Phillips Brooks.[49] Even an incomplete list of the clubs to which he belonged in his adult years places Walker in the company of the economic and political as well as the cultural elite of the eastern seaboard. In Boston, he was at one time or another a member of St. Botolph's Club (president, 1885–95), Union Club, Art Club, Commercial Club, Massachusetts Reform Club, Round Table Club (sociological), Wednesday Evening Club, Thursday Evening Club (vice-president, 1891–97), Saturday Club (at Dr. Holmes's house), Winter Nights Club and University Club. When in Philadelphia, he could enjoy a respite at the Manufacturers' Club; in Washington, DC, at the famous Cosmos Club. In New York City, he belonged to the Century Club, the University Club and the Round Table Dining Club (as distinct from the club of the same name in Boston). Many prominent New Englanders serving in Washington were among Walker's acquaintances, notably James A. Garfield and Senator George F. Hoar, who had taken an interest in the young man's career when Walker was still in his twenties.

Walker fought for four years in the Civil War, moving up through the ranks from sergeant major of the 15th Massachusetts Volunteers in 1861 to brevet brigadier general in March of 1865. He survived many battles, a shrapnel wound which temporarily disabled him, and a trying stint in Confederate captivity at Libby Prison in Richmond, VA. His superior officers held Walker in uniformly high regard for his steadiness under pressure and his logistical abilities. After the war ended, Walker taught classical languages for three years at a preparatory academy, and also tried his hand at journalism in Springfield, MA. His public career was launched in earnest in 1869, when at the age of 29, he was brought on by the respected economist David A. Wells as chief of the Bureau of Statistics in the Treasury Department and special commissioner of US Revenue. There he quickly distinguished himself as able and energetic, and so became an attractive prospect for the Republican controlled appointment to head the Ninth (1870) Census. Enough has been said already in Chapter 2 to make it clear that Walker's performance as superintendent was difficult to quarrel with on grounds of competence or integrity (though his predecessor Joseph Kennedy did so quarrel nevertheless). In order to keep Walker on the federal payroll when funding to pay the superintendent was not renewed by Congress in 1871, he was put in charge of the Bureau of Indian Affairs (BIA), then notorious as a hotbed of corruption. He quickly acquired a grasp of this

---

[49] Solomon, *Ancestors and immigrants*, 70.

new field and earned praise for incorruptibility, publishing a widely read article on "the Indian question" in 1873. While commissioner of the BIA, he continued to supervise the tabulation and publication of census results, his 1874 *Statistical Atlas of the United States* winning international praise. In 1873, Walker accepted a post as professor of political economy at Yale's Sheffield Scientific School, soon making a name for himself in this discipline with an influential critique of the prevailing theory of the determination of wages in a capitalist economy. Subsequent work resulted in an extremely influential textbook on political economy and a treatise on money based on a series of well-regarded lectures on the topic delivered at Johns Hopkins University in 1877. This last got Walker appointed to the 1878 International Monetary Conference held in Brussels (he was asked to represent the US again in 1892 but declined). In 1880, Walker was called away from Yale temporarily to direct the Tenth Census, whose enabling legislation he had helped to draft (see Chapter 2). This work solidified his reputation as one of the preeminent statisticians in the United States. The regard in which he was held is attested by the fact that he was selected to write the entries on "Census" and on "Political geography and statistics" in the ninth edition of the *Encyclopaedia Britannica*.[50]

In 1881, the 41-year-old Walker took over the presidency of the Massachusetts Institute of Technology, which remained his primary professional commitment (despite tempting offers, such as the presidency of Stanford University) until his untimely death in 1897. Characteristically, Walker was soon established as an eminent educational thinker, but continued as well to speak, write and give advice to national and local states on all other subjects with which he was familiar. As befitted someone so prominent, Walker's thoughts appeared in the major magazines of the day, beginning in 1869. All told, he contributed forty articles and reviews on national issues to the top national journals during the course of the next twenty-three years (eight to *Forum*, seven each to *Atlantic* and *Scribners/Century*, five each to *North American Review* and *Princeton Review*, three each to the *Nation* and *Lippincotts* and two to *International Review*). Especially as president of MIT, Walker was invited to speak at numerous inaugurations and special functions, and his remarks were frequently reprinted in eastern newspapers. With such a strong national reputation, international renown was not long in coming, and Walker accumulated a list of prizes, honorary memberships and honorary degrees from monarchs, royal societies and universities throughout Europe.

All of the foregoing accomplishments identify Francis Walker as a charter

---

[50] F.A. Walker, "The Indian question," *North American Review* 66, 239 (1873), 329–388; *Wages: A treatise on wages and the wages class* (New York 1876); *Political economy* (New York, 1888); *Money* (Baltimore, MD, 1878); "Census," in *Encyclopaedia Britannica*, 9th ed., vol. V (New York, 1876), 334–340; "Political geography and statistics," in *Encyclopaedia Britannica*, 9th ed., vol. xxiii (New York, 1888), 818–829.

member of the coterie of best men, and surviving reports depict just the sort of personality one might expect. Although this study is not a biography, some of the specifics of Walker's personality bear mention (and in Chapter 4, analysis). He was noted for resolution, fairness, strength of will, loyalty to cherished causes, people and institutions, and for an abiding interest in the cultivation of such virtues in young adults. Professionally, he was known first and foremost for his impartiality. Indeed he became somewhat of a living symbol of this important virtue, as attested for example by his appointment as chief of all competition judging at the Philadelphia Centennial Exhibition in 1876. Three years earlier, Walker had been appointed by President Grant as one of the commissioners charged with inspecting and approving the annual trial pieces of gold and silver coin at the Philadelphia Mint, a duty he would be called on to perform again.[51] In 1883, he was appointed by the American Association for the Advancement of Science (ASSA) as "Chairman of a Committee to consider the advisability of a recommendation from the Academy of the adoption of the new Standard Time."[52] In short, as Richard Sidwell put it, Walker was a "man of knowledge," a figure to whom the cultural elite looked for answers to the pressing social and economic problems confronting them.[53] Walker's biographer relates that Yale's President Hadley believed Walker "knew more things worth knowing than any other man of his [Hadley's] acquaintance."[54]

This brief *résumé* places Walker among the "old guard" who would soon be eclipsed by academic and bureaucratic specialists. But he was also one of the most prominent participants in the transformation, active both in the national organizations which attempted to further the cause of American science (he was vice-president of the fledgling National Academy of Sciences in 1890, and belonged as well to the ASSA and the AAAS), and in the discipline-specific associations that have given social sciences their modern organizational shape. A longtime member of the ASA, Walker presided over this august group from 1883 until his death in 1897, and was chiefly responsible for transforming it from an institution of the older cultural elite to one whose members often hailed from university departments, as well as for inaugurating regular publications.[55] The young German-trained economists striving to overthrow *laissez-faire* orthodoxy elected Walker to the presidency of the newly formed AEA in 1885, a post he held until 1892. He held honorary memberships also in the American Historical Association and the American Geographical Society. In keeping with these commitments, Walker contributed to the new

---

[51] Boutwell to Walker, Jan. 30, 1873, Francis A. Walker Papers, 1862–1897, Massachusetts Institute of Technology Archives and Special Collections, MC 298, Box 1, Folder 1.

[52] A.A.A.S. to Walker, Oct. 22, 1883, Francis A. Walker papers, 1862–1897, MIT, Box 1, Folder 3.

[53] Sidwell, "The economic doctrines of Francis Amasa Walker," 7–10.

[54] Munroe, *A life of Francis Amasa Walker*, 152.

[55] Koren, "The American Statistical Association," 12–13.

specialized journals, most often to the *Quarterly Journal of Economics* but also to *Political Science Quarterly* and the *Journal of Political Economy*.

He became an advocate as well for specialization in the institutional structure of higher education, believing the cause of science to be aided by the existence of technical and scientific schools such as MIT and by the modern departmental form. He was a frequent correspondent (and ally) of Daniel Coit Gilman, the innovative president of Johns Hopkins who had done so much to remake the landscape of American higher education. In short, Walker ended up as thoroughly immersed in the new culture of expertise as he was in the old culture revolving around the traditional northeastern elite. He did not himself become a specialist in the fullest sense (his articles in specialized journals remained general and accessible to a wide audience). But he wholeheartedly embraced the new array of institutions and subject positions associated with expertise. He had a prominent place in both the early and the late phases of evolving governmental culture, and an unusually broad field of practical and intellectual competence.

Straddling the late nineteenth-century American governmental discursive formation in so many senses, Walker is eminently qualified to serve as a representative of governmental subjectivity. At the very least, any student of national-scale governmentality confronted with a career like his could hardly resist wondering what he thought. I will lay out his program in great detail in Part II, and use it as a platform for exploring the spatial politics which I contend are necessarily central to national governmentality in general. I will trace the outlines of a governmental spatial politics in Walker's involvement with all three "moments" of the cycle of social control ("observation," "judgment" and "enforcement" or "regulation"). But again, the substance of Walker's (or anyone else's) governmental program will be woven from contemporary cultural threads reaching far beyond relatively narrow issues of spatial politics. As I pursue Walker's symptomatic spatial politics in the latter half of the book, it will also become clear that gender issues, in particular an overriding concern with the threat posed by modern industrial life to American manhood, motivated his governmental program in fundamental ways.

I could treat this as an idiosyncratic curiosity, and deal with it only in passing. After all, other contemporaries of Walker's showed different emphases in their discussions of social order. But the feminist scholarship I discussed above strongly suggests that this would be a mistake. In Walker's case, I believe the concern with manhood links the rise of governmentality to another general and systematic dynamic in American culture. To put my hypothesis in schematic terms, the strain suffered by Walker's generation of early governmental thinkers as they attempted to retain individual mastery of an ever more unmanageable field of thought and action was experienced as a threat to the older conceptions of manhood through which they understood their identities and careers. This predisposed at least some of them to project the issue of manhood into their analyses of the body politic at a basic level. In so doing, men like Walker demonstrated how a modern governmental program and its underlying spatial politics could be thoroughly gendered.

I hope in chapter 4 to provide a context for understanding Walker's experience of the strains of governmentality, and also to contribute to a more nuanced understanding of American manhood. I will contend that Walker's experience raises some interesting issues regarding the transition between manhood ideals in the age of relatively unfettered capitalism and manhood ideals as they changed to accommodate the "Progressive" era of specialization and bureaucracy. At the end of this detour, all the pieces will be in place for understanding the details of Walker's governmental program in Part II.

4

American manhood and the strains of governmental subjectivity

Francis Walker (Figure 1) lived during a period of great change, and, again, the rise of governmental programs like his can be understood as an important part of what Robert Wiebe has termed "the search for order."[1] The changes which provoked this search have been well rehearsed: industrialization and attendant class conflict, urbanization, immigration, the movement of women into public life, the closing of the western frontier. An older generation of historians (of which Wiebe is one) tended to view these changes as challenges to "American society"; but more recently, historians of class, feminist historians and historians of race and ethnicity have begun to ask a more pointed question: "challenges to whom, precisely?"[2] In this light, the crisis ceases to appear as a sort of "adaptation" challenge for an essentially harmonious "American people," and becomes instead a matter of cultural and social politics. Gender is a primary dimension of cultural and social reality, as Joan W. Scott explains so succinctly:

> Since all institutions employ some divisions of labor, since the structures of many institutions are premised on sexual divisions of labor (even if such divisions exclude one sex or the other), since references to the body often legitimize the forms institutions take, gender is, in fact, an aspect of social organization generally.[3]

It is not surprising that with the benefit of such basic insights, historians have recently been able to document a serious "crisis of manhood" spanning the late nineteenth century. In a sense, the crisis of manhood was nothing other than the general "social" crisis of the period manifested at the level of the needs, desires and insecurities of those for whom gender was a primary way to understand their positions with respect to the larger context.

E. Anthony Rotundo notes that "so many of our institutions have men's

---

[1] R. Wiebe, *The search for order, 1877–1920* (New York, 1967), 12.
[2] See, for example, D. Montgomery, *The fall of the house of labor* (New York, 1987); J. W. Scott, *Gender and the politics of history* (New York, 1985); A. Saxton, *The rise and fall of the white republic* (New York, 1990).   [3] Scott, *Gender and the politics of history*, 6.

needs and values built into their foundations, so many of our habits of thought were formed by male views at specific points in historical time, that we must understand gender in its historical dimension to understand our ideas and institutions."[4] Governmentality, understood as one of our "habits of thought," is no exception. Among his many other distinctions, Francis Walker played a prominent role as an exemplar of the masculine ideal in pursuit of which many American men encountered the crisis of manhood. But even Walker's ample manhood ultimately failed him in his attempts to master a rudimentary governmental system. His immersion in the culture of manhood, and the particular way in which that culture failed him, combined to produce a definite gendering of the approach to governmentality Walker bequeathed to his successors. In what follows, I will first briefly sample the secondary literature in order to lay out some of the basic dimensions that structured thinking about manhood in the late nineteenth century, then show how thoroughly immersed Francis Walker was in these structures of thought, and finally, show how his attempt to ground a manhood sufficient for governmental mastery of the national territory failed.

Manhood and its discontents in late nineteenth-century America

The root issues confronting men in the late nineteenth century concerned the effects of loss of control. To translate the "general" features of social crisis noted above into the terms of individual male experience, the accelerating scale, pace and interconnectedness of material life brought a generally felt loss of control over the conditions of existence; the erosion of traditional race and gender hierarchies brought a loss of control by native-born white men over other groups (women, workers and blacks); the stirrings of class conflict threatened a loss of control by the relatively wealthy over the distribution of resources (political and economic power) among groups; the old money elite steadily lost economic and political ground to the *nouveaux riches* created by manufacturing capital; and the rise of more or less radical recipes for change constituted a loss of control by cultural and political elites (old and new) over the definition of legitimate social order. Finally, a new perception that land and other material resources were finite relative to the numbers of people wanting access to them forced more men at all levels into a competitive relationship with other men, which threatened to intensify the experience of loss of control still further.

Clearly, loss of control meant a range of different things to men in different social, economic and cultural subject positions. The traditional category of "class" is inadequate to capture the most basic divisions; perhaps the more

---

[4] E. A. Rotundo, *American manhood: Transformations in masculinity from the Revolution to the modern era* (New York, 1993), 8–9.

recent term "stakeholder" defines the issue more accurately. The various male subject positions can be separated into two "camps" according to whether the men who occupied them accepted the basic outlines of the emerging industrial system as a base from which to continue pursuing individual goals and/or improving American society. Thus the "stakeholders" would include not only industrial capitalists and the emerging professionals but also most elite social reformers, as well as most clergy and some (though not all) farmers. Some stakeholders deeply distrusted the new political economy, but still felt that it was better to attempt to improve it than to abandon it and risk more drastic transformations of social order. "Non-stakeholders" might also reconcile themselves to the new order, but generally had compelling reasons at least to question its legitimacy. For those men less committed to preserving the out-lines of the social order, loss of control revolved mostly around the threaten-ing aspects of the change from yeoman farmer or artisan status to that of landless tenant or proletarian.

I will argue below that this distinction between stakeholders and non-stake-holders helps us to understand the dynamics of the crisis of manhood, but it should be emphasized that men on both sides of the divide were caught up in the crisis. Economic insecurity of course weighed more heavily on non-stake-holders, but many stakeholders also came to understand their situation as pre-carious. In an important sense, the construction of a crisis of manhood represented an attempt to link the interest of stakeholding and non-stake-holding men *as men*, an attempt to forge what Dana Nelson terms "national manhood."[5] It was a crisis addressed to all men, but a crisis whose terms and solutions were constructed by those with an interest in preserving at least the basic features of the emerging industrial society.

With these issues of authorship and audience straightened out, it is possible to make better sense of how the late nineteenth-century crisis fits into recent historical accounts of American manhood. Both Michael Kimmel, in his *Manhood in America*, and E. Anthony Rotundo, in *American manhood*, skirt these difficulties somewhat, though for different reasons. Rotundo's study is limited in scope to the cultural elite of the northeast, and concentrates mainly on changes in this group's self-understanding through the nineteenth century; Kimmel tends to treat "American manhood" as an already unified category even in the early nineteenth century, thus taking for granted the success of what Judy Hilkey shows was at least in part a deliberate political project.[6] Nevertheless, both Rotundo and Kimmel provide useful accounts of the rela-tionships between nineteenth-century social changes, the experience of these

---

[5] D. Nelson, *National manhood: Capitalist citizenship and the imagined fraternity of white men* (Durham, NC, 1998).

[6] Rotundo, *American manhood*; M. Kimmel, *Manhood in America: A cultural history* (New York, 1996); J. Hilkey, *Character is capital: Success manuals and manhood in Gilded Age America* (Chapel Hill, NC, 1997).

changes as threats to manhood, and the construction of new manhood ideals (as well as the formulation of new gender-based social strategies) through which to meet these threats.

Kimmel tracks the crucial transformation which occurred in the early nineteenth century, when the "self-made man" rose to prominence as a cultural ideal with the decline of landownership (the "Genteel Patriarch") and of "the self-possession of the independent artisan, shopkeeper or farmer (the Heroic Artisan)" as secure anchors for male identity. Ideals of American manhood ever since that time have revolved around the self-made man, whose overarching concern as a man is to prove his manhood.[7] The crisis of manhood experienced in the late nineteenth century was essentially a crisis of the self-made man. It derived from the fact that while self-made, outwardly visible success was supposed to underpin masculinity, it was rapidly becoming less available to most men. Though stakeholders had much better prospects for success (in the professions and government as well as in business), it was increasingly difficult even for them to achieve self-made success.[8]

The core substantive ingredients of self-made manhood, recommended by success-manual writers and social commentators, remained stable throughout the period from the Civil War to World War I (and has arguably not altered much since).[9] American manhood was ideally built on two things: character and will-power. Character was composed of very traditional virtues: a strong sense of duty, courage and bravery, steadiness, self-control, self-sacrifice, loyalty and industriousness.[10] This complex of virtues was supposed to comprise a sort of civilizing harness for the prodigious will-power, the motive force or engine of the vigorous man. A man endowed with sufficient "push" or assertiveness would benefit most if he could control, channel, steer and conserve this energy for use at the proper moments and in the proper activities.[11] The idea of "reserve force" was crucial, since will-power was a finite resource. It would ill behoove a man to be caught in a time of exigency without the surplus vigor necessary to overcome unexpected obstacles. The proper balance of character and will-power would allow a man not only to cope with but actually to thrive on the demanding struggles of modern life.

The sexual imagery in all of this is thinly veiled, and indeed many connections were drawn between the husbanding of manly energy and the conservation of sperm. It was of paramount importance for men to use their "nerve force" wisely and constructively, just as it was necessary to avoid spilling seed

[7] Kimmel, *Manhood in America*, 9–10. See also A. McLaren, *The trials of masculinity: Policing sexual boundaries, 1870–1930* (Chicago, 1997), 34. McLaren shows, through an examination of court cases, that proving and policing manhood was an important cultural issue in all Euro-American nations in the late nineteenth century.   [8] Rotundo, *American manhood*.

[9] McLaren, *The trials of masculinity*, 131. This stability explains Kimmel's stress on the "self-made man" as the dominant ideal from the early nineteenth century until the present day.

[10] Rotundo, *American manhood*, 12–13; Kimmel, *Manhood in America*, 112; Hilkey, *Character is capital*, 127.   [11] Hilkey, *Character is capital*, 145–149.

in unproductive activities other than procreative sexual intercourse.[12] Angus McLaren chronicles the judicial enforcement of this "spermatic economy" in Canada, France and England as well as in the United States during the nineteenth century. In all of these countries, there was intense concern both about the debilitating effects industrialization might have on the individual and about decreasing national fertility rates. The two were of course linked, as either the loss of nerve force through what we would today call "stress" or the careless spending of sperm on the part of individual men (or both) were thought to be behind the alarming drop in population growth rates.[13] Such was the thinking behind Teddy Roosevelt's campaign to head off "race suicide" as native-born white American fertility lagged that of European immigrants at the turn of the twentieth century; and such was the mindset within which Francis Walker's understanding of the American social body was constructed (see Chapter 6).

Tom Lutz argues persuasively that much of what was thought and said about the threats to manhood and the larger social order during the late nineteenth and early twentieth century can be interpreted as part of the "discourse of neurasthenia."[14] This is confirmed circumstantially by the fact that no historian of American manhood fails to dwell on the topic of neurasthenia at some length. Neurasthenia was a name given to a broad range of ailments thought to express "nervous exhaustion" in both men and women. Lutz's analysis strongly suggests that it was an entirely "cultural" disease, that it was not a physiological phenomenon at all, except in a superficial psychosomatic sense. One of the symptoms of its "cultural" character was the fact that a diagnosis of neurasthenia brought diametrically opposite prescriptions for male and female sufferers. The cause of strain was often the same for both sexes: thinking too much (neurasthenia was overwhelmingly an ailment of the "thinking classes"). The neurasthenic woman was almost invariably given rest cure; the neurasthenic man, by contrast, was usually thought to need vigorous exercise and fresh air. Thus "nerve force" was not understood as a leveling factor that brought men and women onto the same plane as equivalently constituted human beings. Both sexes could suffer from thinking too much, but the answer for men was exercise, while for women it was vacant repose. Neurasthenic discourse, then, was thoroughly articulated with a gender ideology stressing fundamental differences between the two sexes. Vaguely and inconsistently conceived though it was, neurasthenia was generally understood to be related to the accelerated pace of American life in particular. For Lutz, the important thing about neurasthenia is not so much its details, or its coherence as a medical concept, but rather the ubiquity of its language among the economic and cultural elite of the late nineteenth and early twentieth

---

[12] *Ibid.*, 148.   [13] McLaren, *The trials of masculinity*, 155, 179.
[14] T. Lutz, *American nervousness, 1903* (Ithaca, NY, 1991).

centuries. This ubiquity signaled a widespread acceptance of underlying assumptions about "nerve force" as a finite resource to be augmented whenever possible and spent only when necessary. This was a key component of the discourse about manhood.

By using some of these categories with a heightened awareness of who was trying to persuade whom of what, we can characterize the late nineteenth-century crisis of American manhood as follows: faced with a higher level of personal insecurity, as well as with uncomfortably serious threats to social order, stakeholding men felt the need to bolster (and in some ways transform) the ideal of the self-made man by inventing and encouraging coping strategies for themselves and for non-stakeholders that were compatible with the basic outlines of American capitalism and the culture surrounding it. Some of these strategies (the trappings of Rotundo's "passionate manhood") revolved around self-government and the individual pursuit of manhood ideals, while others were collective and directed at various "others." Some collective strategies (for example, nativism and gender discrimination) were already part and parcel of the cultures of (white) workers and yeoman farmers, and in such cases the goal of stakeholders was merely to use them to forge male solidarity in ways that would avoid drastic transformations of the socioeconomic system. It is crucial to stress here that all of these "strategies" were at best only imperfectly understood as such by their proponents, and that the hint of functionalism suggested by calling them "strategies" is a symptom of retrospective analysis. Nevertheless, they were not pursued in total ignorance of their larger implications for social order, and it is possible to understand a great deal by treating them as strategies. As pressures on white men intensified toward the end of the nineteenth century, a variety of solutions were offered by stakeholders, most urgently to workers and poorer yeoman farmers. For these non-stakeholders, the ideal of self-made manhood was most palpably questionable. In the cities, the labor movement championed an alternative manhood ideal based on the principle of solidarity, and organizations such as the Knights of Labor seemed during the 1880s to demonstrate real promise in this alternative. At the same time, in the rural United States, grangers and Farmers' Alliance lecturers were busy educating the yeomanry on economic principles that cast the emerging corporate economy in a highly negative light.[15] To counter these threats, to present the masses of striving men with an attractive and workable image of American manhood, writers of (wildly popular) success manuals codified a number of arguments also prevalent in newspaper and magazine discussions of manhood. The hope was that (partly through their influence), class and other volatile distinctions would be rendered unimportant.

The key problem for success-manual authors (and for stakeholders in

---

[15] Hilkey, *Character is capital*, 143; Kimmel, *Manhood in America*, 107–108.

general) was how to de-couple successful manhood from material success in the minds of those less likely to achieve the latter. One approach was to style manhood in and of itself as success. This allowed dignity to coexist with poverty or lack of social mobility, and also licensed the claim that economic "classes" as conventionally understood did not actually exist. In reality, instead of two opposed "classes" in the conventional sense, there was a community of hard-working, productive *men*, linked together across divisions of wealth or power, and united against the unmanly or dependent.[16] As we shall see in Chapter 6, Francis Walker subscribed to this view, and to the explicit denigration of various "others" that went along with it. Women should stay in the home because they were unfit for the scrap and scramble of public life. Immigrants lacked the manly fibre necessary to compete with full-blooded American men, who if left to themselves in the public arena would perform prodigies of wealth-making in the course of honorable combat for wages and profit. Social Darwinism was often of service in justifying the inequalities always implicit in such glorifications of competition.[17] As Richard Hofstadter has shown, the rhetoric of "survival of the fittest" saturated the discourses of late nineteenth-century American culture, forming a central theme in many areas of social commentary.[18]

Another argument common in advice to the average man portrayed poverty as a desirable condition of the possibility for proving one's manhood. Adversity was an opportunity which actually elevated the lowly above the well-to-do. Those men unfortunate enough to be born into comfortable circumstances would find it difficult to encounter the obstacles necessary for the forging of strong masculinity.[19] Like the other arguments, this one ideally functioned to reconcile working-class or farming men to a more modest degree of economic independence.

Stakeholders had to make some adjustments in their ideas of their own independence, as well. Businessmen and merchants organized their fears through the image of the vulnerable competitive position of the entrepeneur. This image also became influential in the self-understandings of members of the eastern political and cultural elites such as Walker who were not, strictly speaking, entrepeneurs. For both groups, as for non-stakeholders, there was growing pressure to prove one's manhood through outwardly visible achievements. Rotundo sees the adjustments to the meaning of independence amounting to a new manhood ideal, the "passionate man." Around the turn of the twentieth century, stakeholding men came increasingly to value the "male passions" of aggression, assertiveness, strength and even boyish impulsiveness as positive expressions of self rather than suspicious impulses to be

---

[16] Hilkey, *Character is capital*, 91–96.   [17] Kimmel, *Manhood in America*, 84–100.

[18] R. Hoftsadter, *Social Darwinism in American thought* (Boston, MA, 1992 [1944]).

[19] Hilkey, *Character is capital*, 129–131; K. Townshend, *Manhood at Harvard: William James and others* (New York, 1996), 171.

controlled.[20] For the passionate man, not only had the male passions become ends in themselves and points of pride, but just as importantly, independence changed its meaning from self-reliance to freedom from restraints on self-expression. This change dovetailed with a larger transformation of middle-class consciousness toward an understanding of identity in terms of individual desires and their satisfaction through consumption.

An important component of self-expression was physical culture, and it was around the turn of the twentieth century that body-building, boxing and other such outward displays gained wide popularity. Competitive sports were held up as a crucial substitute for the kind of frontier existence no longer available to harden American manhood.[21] Intellectuality faded as a feature of manliness, in favor of an ideal centered more exclusively on strength and "hardness" (cementing a general cultural complicity between masculinity and distrust of excessive thinking that has endured, arguably with very destructive effects, through the twentieth century[22]). Not coincidentally, just as masculinity became a matter of personal expression, homosexuality came in for far more intense stigmatization. It was around the turn of the twentieth century that homosexuality first came to be perceived as a matter of inherent personal identity rather than of decisions, acts and indulgences of otherwise "normal" men.[23] Men increasingly feared and rejected homosexuals not only out of fear of "the woman within" but also out of concern with another of the large-scale social issues occupying late nineteenth-century elites throughout Europe and North America: the declining birth-rates of industrializing, urbanizing nations. Walker was only one among many who would make these links as he pondered social problems (see Chapter 6).

Finally, men at all levels of the socioeconomic hierarchy avidly formed and joined homosocial organizations during this period. Fraternal lodges sprang up everywhere, and college fraternities had their heyday in the years around

---

[20] Rotundo, *American manhood*, 2–7.

[21] R. J. Park, "Biological thought, athletics and the formation of a 'man of character': 1830–1900," in J. A. Mangan and J. Walvin (eds.), *Manliness and morality: Middle-class masculinity in Britain and America, 1800–1940* (New York, 1987), 7–34.

[22] Richard Hofstadter's classic *Anti-intellectualism in American life* (New York, 1962) is extremely interesting to read through the lenses of more recent scholarship on gender. The saturation of Hofstadter's primary sources with the language of manhood, which went unremarked at the time, amply illustrates one of Rotundo's more profound observations:

> In the end, men and women alike are harmed [by inherited ideas about manhood], because these symbols of right and wrong manhood have also become lodged in our political consciousness and in the decision-making culture of our great institutions. These symbols make certain choices automatically less acceptable, and in doing so they impoverish the process by which policy is made. We are biased in favor of options we consider the tough ones and against those we see as tender; we value toughness as an end in itself. We are disabled in choosing the wise risk from the unwise, and tend to value risk as its own form of good. (291)

[23] *Ibid.*, 274–279.

1900. Fraternal orders "attempted to create a domestic sanctuary outside the home – a place where men might experience fellowship and intimacy without the feminizing influence of women."[24] This was a mode of escape and exclusion that helped men deal with the growing presence of women in the traditionally male public domain. As Chapters 6 and 7 will show, nativism and immigration restriction movements, which involved both workers and elites and resulted in the Chinese Exclusion Act of 1882 and the Immigration Restriction Act of 1924, were saturated with a rhetoric of manhood that painted immigrants as grave threats to American virility. Through most of this period as well, Indians, blacks and women were denied citizenship rights, either formally or informally, in part through a rhetoric of threats to manhood.

All these new twists on the images and practices of manhood were more or less consciously intended to unite native-born white American men in a community of interest, to put them in the best possible position to cope with the foreclosure of widespread upward mobility without resorting to class struggle, and to counter the surge of immigrants into American life, the appearance of women in public spaces and the generally faster pace accompanying the industrializing world. However well they succeeded, though, none of these strategies eliminated the pressure on individual men to prove their manhood.

To sum up this brief survey, stakeholding white, native-born American men met the challenges of the late nineteenth century by attempting to exercise the virtues making up "character," and by using their will-power judiciously (in accordance with neurasthenic economy). Bringing these resources into play, it was hoped, would allow "real men" to *make* their own success, regardless of how many circumstances conspired against the possibility. In the aggregate, the strivings of all these men would push America onward toward greatness. But in so far as this scenario was becoming less plausible, a combination of supplementary strategies was in order. Escape into carefully preserved homosocial settings; exclusion of blacks, immigrants and women from as many aspects of public life as possible; cultivation of their own manhood as personal expression; and propaganda directed toward non-stakeholders and geared to head off class conflict, all helped these men construct a more or less successful national culture of manhood.[25]

With the collapse of populism and industrial unionism in the late nineteenth century, and the ascendancy of collaborationist, reform-oriented unionism by the end of World War I; with the consolidation of segregation in the South and nativism in the North; and with the (temporary) muting of feminism after women gained the vote, the most significant challenges and alternatives to this model of masculinity were seriously undermined. Men who had

---

[24] Kimmel, *Manhood in America*, 173; Rotundo, *American manhood*, 200–203.
[25] See Nelson, *National Manhood* for a searching analysis of the origins of this project in the Constitutional era.

been "non-stakeholders" found it increasingly difficult (if indeed they tried at all) to distance themselves from the rewards (and the considerable demands) of the "self-made" ideal man. The "crisis" of self-made manhood never entirely disappeared, but it was muted somewhat, and decoupled from the possibility of drastic social upheavals.

Francis Walker as an exemplar of manhood

Francis Walker, as I have already suggested in a number of places, was thoroughly immersed in what might be called the "cult of true manhood." His biography, completed by James P. Munroe (a former student) in 1923, is at one level nothing more than an extended paean to manhood.[26] The first half of the book is taken up with an exhaustive account of the four years comprising Walker's Civil War experience, and is entirely given over to the interpretation of that experience in terms of manhood. The rest of the book recounts his prolific public career as a straightforward continuation of the military struggle, prosecuted in precisely the same way as a battlefield campaign. Munroe's prose is saturated with manly rhetoric throughout (Walker is styled a "virile" thinker, a "vigorous" man, etc.), and a wealth of anecdotes illustrate Walker's possession of all the virtues that go to make up a manly "character." A sampling should impart the general flavor of the life Munroe constructs.

Despite offers of a "leg up" from influential Massachusetts patrons, the 21-year-old Walker insisted on entering the Union Army at the lowly rank of sergeant major (thereby acquiring the opportunity to "make himself," which he quickly did). He wrote home of the glories of battle ("Ah, Kate, don't you wish you were a man and could fight?"), and railed against the "ministers and lawyers and merchants" criticizing military movements from a safe distance. According to Munroe, he "was a born soldier, loving the struggle of mind against mind, of strength against strength, just as throughout his life he tingled with joy at the clash of the football field." Wounded in the hand by shrapnel, he was sent home, and his hand subjected to one of the first X-rays taken in the United States. The doctor inadvertently gave Walker a strong electric shock, and was moved to comment, by Walker's stoic reaction, that "there are times in life when military discipline shows itself." Deprived of rations at one point by a logistical mistake, Walker simply ceased to eat for three days rather than "beg" from his comrades. As he rose through the ranks, Walker displayed a willingness to take up whatever work was at hand, regardless of whether it "fitted his station," and to continue at it until the task at hand was complete. He was the model of obedience to his superior officers, and fiercely loyal both to officers and to troops.[27]

---

[26] J. Munroe, *A life of Francis Amasa Walker* (New York, 1923).
[27] *Ibid.*, 31–36, 36–37, 48, 65–66, 71–72, 101.

Munroe sees a close parallel between Walker's character and that of General Winfield Scott Hancock, under whom Walker served and whose biography he later wrote. The virtues Walker ascribed to Hancock can serve as a template through which to understand Munroe's interpretation of Walker:

Hancock was not a man of lofty intellectuality. He had courage – fiery, enthusiastic courage; positive, active, unfaltering loyalty to country and comrade; he had industry beyond measure; the ambition that stirs to do great deeds, and be worthy of high promotion; above all an unrest while anything remained to be done; a dissatisfaction with what was incomplete; a repugnance at all that was slovenly, clumsy, coarse, or half made up.[28]

The "engine" of Walker's manhood, his "will power," is also well attested in the biography. The countless references to his "strenuous efforts" in practically every dimension of public life imply an almost bottomless fund of nerve force. One of its more spectacular manifestations was a "hasty, almost ungovernable temper," and the biographer recalls with some awe being present when Walker was driven to "Herculean efforts" to control this temper.[29] This temper is worth more extended consideration. Peter Stearns has recently chronicled a change in attitudes toward male anger in America over the course of the nineteenth century, arguing that earlier negative perceptions of anger were superseded toward the end of the century by a new ambivalence.[30] This ambivalence stemmed from the increasingly positive value placed on aggressive assertion, which balanced the awareness of its still obvious destructive potential. G. Stanley Hall (a prominent psychologist, first president of Clark University, and a contemporary of Walker's) claimed that "a choleric vein gives zest and force to all acts."[31] Consistently, Hall "recommended boxing as a perfect means of training adolescent males in anger control, for it channelled the emotion away from unwarranted targets without quelling it."[32] In keeping with this analysis, the volcanic anger with which Walker wrestled would have been taken by contemporaries as a healthy expression of manly vigor.

The two basic modes of expression of will-power, aggressive outward assertion and the effort of self-control necessary to rein it in, are displayed to great effect throughout Munroe's life of Walker. Walker himself clearly understood his experiences in such terms, and felt himself to possess unusually potent nerve force. For example, he wrote in 1885 to the wife of the great British economist Alfred Marshall that "I wish I could lend your husband a little of my superabundant vitality to enable him to do his work at more leisure and with more pleasure than his delicate health allows."[33] Among other things, this

---

[28] *Ibid.*, 266; F. Walker, *General Hancock* (New York, 1895), 29.
[29] Munroe, *A life of Francis Amasa Walker*, 25.
[30] P. Stearns, "Men, boys and anger in American society, 1860–1940," in Mangan and Walvin, *Manliness and morality*, 75–91.   [31] Quoted in *ibid.*, 81.   [32] *Ibid.*, 84.
[33] Quoted in Munroe, *A life of Francis Amasa Walker*, 308.

"vitality" produced five children over the course of the years (at unrecorded expense to the vitality of his wife).

Another feature of Munroe's construction of Walker perfectly in line with the "cult of true manhood" is the almost total absence of references to women, most glaringly to the woman without whom Francis Walker would not have been able to achieve all that he did. We learn on page 104 that "on August 16, 1865, he married Miss Exene Stoughton, eldest daughter of Mr. Timothy Morgan Stoughton, of Gill, Massachusetts." Introduced in a single sentence, this woman then disappears for the remainder of the biography, appearing only once or twice more in oblique references. There is of course a great danger in devoting significant space to the figure who most clearly calls into question the image of the "self-made" Francis Walker.[34] We shall see in Chapter 5 that, at times, Walker wrote about the domestic sphere of reproduction as though it were almost an alien planet which he personally had never visited.

It was not merely Walker and his biographer who saw in him an exemplar of manhood. Many of his contemporaries undoubtedly saw him in similar terms; two examples should suffice to illustrate this. In 1875, Charles Francis Adams (of the famous Massachusetts Adams) wrote to Walker explaining his own decision not to attempt to organize and run the Massachusetts exhibition at the upcoming Philadelphia Centennial. In accepting Adams's decision, the governor of Massachusetts "kindly intimated . . . that he would pay some consideration to anything I suggested, and I told him that a *man* was wanted – no more and no less. I then suggested you as the best man available, and the suggestion was well-received."[35] As it happened, Walker was appointed chief of the Bureau of Awards for the whole centennial exposition, and thus could not represent Massachusetts. In recognition of his work at the close of the exposition, his coadjutors drew up a document in October of 1876 which included a mock-formal award:

Department – Special
    Exhibitor – The United States of America
        Group – Genus Homo
        Class – Number 1
        Object – A Thoroughly Developed Man, Chief of the Bureau of Awards

Beyond any shadow of a doubt, the "cult of true manhood" was one of the primary, probably *the* most important, conceptual structure through which Walker understood himself and his work.

Recent scholars, touching on different points, have by and large accepted (and in the aggregate reinforced) the construction of Walker as an ideal man for the modern America then emerging. For example, Kimmel, Rotundo and Kim Townshend (in *Manhood at Harvard: William James and others*) all

---

[34] See Hilkey, *Character is capital*, 163.   [35] Quoted in *ibid.*, 168.

highlight Walker's *Phi Beta Kappa* speech at Harvard in 1893 as among the strongest statements of the age on the value of athletics in cultivating manhood. In this context, Townshend goes so far as to label Walker "the ideal spokesman for his era."[36] Historian of social science Dorothy Ross places Walker at the vanguard of a movement to invest early American social science with an explicitly masculine empirical realism; cultural historian David Shi makes the connection between Walker's manly fact-facing and the larger move toward realism in the post-Civil War decades.[37] Historian of economic thought Eric Roll makes a more oblique (and very interesting) connection: "Walker worked in a number of fields in all of which he distinguished himself by a considerable energy and by the vigorous espousal of definite views."[38] Like Munroe, these scholars paint a picture not of "lofty" but rather of force-ful intellect. Townshend gets to the nub of the matter when he notes the con-sistency in the fact that for Walker "the steps from war to science to the economy seemed naturally to lead to 'college athletics,' even on an occasion honoring academic achievement."[39] This link between war, manhood and science is the chain out of which Walker attempted to forge a figure of manhood equal to the program of governmental mastery he set for himself. In the next section, I explore the nature and limits of Walker's military manhood ideal, and show how its failure as a model for governmental subjec-tivity reinforced Walker's preoccupation with manhood as the basis of social order.

Military manhood and governmental subjectivity

There is obviously a strong connection between the dimensions of the strug-gle to achieve manhood in late nineteenth-century America and the military life.[40] Most of the traditional virtues understood to make up character are among the "martial" virtues, which should come as no surprise in light of the almost universal characterization of life as a "battle." Though the context for "competition" might be the world of business and commerce, its essence was understood to be *fighting* (thus the definition of will-power in terms of aggres-sion and self-assertion). Kimmel, Rotundo and others make it clear that mil-itary experience held a privileged place in the culture of manhood.

The great historical reference point for the martial aspects of manhood

---

[36] Kimmel, *Manhood in America*, 137; Rotundo, *American manhood*, 240; Townshend, *Manhood at Harvard*, 98.

[37] D. Ross, *The origins of American social science* (New York, 1991), 59; D. Shi, *Facing facts: Realism in American thought and culture, 1850–1920* (New York, 1995), 72.

[38] E. Roll, *A history of economic thought*, 4th ed. (New York, 1992), 385. This, presumably, is how a man doing intellectual work reconciles himself to an anti-intellectual cult of manhood (see footnote 11 above).   [39] Townshend, *Manhood at Harvard*, 98.

[40] D. J. Mrozek, "The habit of victory: The American military and the cult of manliness," in Mangan and Walvin, *Manliness and morality*, 221.

until just after Walker's death in 1897 was the Civil War. This conflict simultaneously certified the manhood of its veterans and helped to militarize ideals of masculinity in general. Munroe placed great stress upon the role played by the war in forming Walker's manhood:

More than anything else, the war matured Walker, as it did most of his contemporaries, far beyond his years. He went into the conflict, in the summer of 1861, a boy, and, as he frequently declared, a very "green" boy. He came out of it, less than four years later, a man, with the poise, judgment and sense of genuine values of middle-age. Before reaching twenty-five he had met with responsibilities seldom encountered, under normal conditions, by mature men of large affairs; in the years when most youths are quaffing thoughtlessly the wine of life, he had been daily face to face with death . . . Trained in a school of the hardest kind of work, sustained and exhausting labor had for him no terrors. Educated early in the bewildering duties of a staff position, details were never too great or too complex for his eager mastery. To the war, therefore, Francis Walker owed in great measure that self-command, that seriousness of outlook, that love of hard, purposeful work which not only enabled him to accomplish such prodigious labors for the public welfare, but fitted him to deal so wisely with young men and to inspire them so loftily.[41]

Most crucially for my purposes here, the changes wrought in the cult of masculinity by the Civil War ushered in a new approach to the great problems and issues facing an industrializing society. George M. Fredrickson documents the emergence of a new style of social and economic reform, characterized by a new skepticism toward (implicitly feminine) "lofty ideals" in favor of (explicitly masculine) hard empirical facts, and personified by Francis Walker, Charles Francis Adams and Oliver Wendell Holmes.[42] All three drew from their wartime experience as officers a firm belief in vigorous, no-nonsense, fact-based methods of coping with the pathologies of Gilded Age American life. The facility they gained in handling great masses of information and in carrying out complex organizational and logistical tasks gave them tools from which they believed civilian society could benefit. Adams would apply what he had learned to the study (and the running) of railroad systems, the largest, most complex business organizations in existence before the 1890s; Walker would apply his experience as assistant adjutant general, which centered on the collection, organization and dissemination of information and orders, to the construction and tireless advocacy of a rudimentary system of national governmentality.[43] Morton Keller makes a similar connection, noting that the Civil War "schooled a generation in the uses of government power to effect social change." Not surprisingly, he, like Fredrickson, explicitly includes Walker (this time along with Carroll D. Wright and David A. Wells) among the "economist-bureaucrats . . . who emerged from the war with a strong belief

---

[41] Munroe, *A life of Francis Amasa Walker*, 103–104.
[42] G. M. Fredrickson, *The inner Civil War: Northern intellectuals and the crisis of the Union* (New York, 1965), 201–209.   [43] Munroe, *A life of Francis Amasa Walker*, 62.

in the capacity of social science and active government to effect social change."[44]

The Civil War not only provided Walker with organizational experience, but probably also encouraged his commitment to forging a unified American nation. Of course this was one of the explicit goals of the Union war effort, but even more generally, according to Douglas Mrozek, "at virtually any time in its history, the US military have been more likely than civilians to envision a unitary culture and society."[45] The basic wartime emphasis on order as the foundation of success would have been an inescapable lesson for all these men.

Theodore Roosevelt is the figure most often associated with the attempt to link martial virtues and national state power, and the culture of scientifically grounded reform associated with the Progressive Era was certainly suffused with the language of "vigor."[46] But this later manhood ideal, stressing duty and service to the state, already took for granted large-scale organizational divisions of labor and the culture of expertise. Thirty years earlier, Francis Walker had plunged into his civilian career determined to make the most of his war-forged manhood, but in a very different way. Instead of devotion to the upright performance of duty within modest and precisely circumscribed organizational roles, Walker was animated by the pursuit of mastery. The point is not that he wanted to *be* the state or any part of it, but rather that, unlike the men of Roosevelt's age, Walker could not conceive of large-scale pieces of social policy systems whose workings were beyond the comprehension of any one individual. He still assumed that even the vast cycle of social control he was striving to put in place, whereby national information-gathering would be linked through informed political-economic theory to national efforts at regulation, could only work if the men in charge of its major elements (for example, the superintendant of the census) each had the "big picture" as well as exhaustively detailed knowledge of their respective pieces. No matter how complex, systems of government could only operate through the rationality embodied in the men who ran them. In keeping with the culture of "impartiality" discussed in Chapter 3, Walker lacked a handy way of conceiving governmental structures as based primarily on vast systems of standardized procedures, regulations and positions, and only secondarily on the people staffing those positions.

In other words, the military manhood ideal Walker brought to his govern-

---

[44] M. Keller, *Affairs of state: Public life in late nineteenth century America* (Cambridge, MA, 1977), 123–124. An important dissenting voice on this issue is T. J. Jackson Lears, whose *No place of grace: Antimodernism and the transformation of American culture, 1880–1920* (Chicago, 1983) portrays militarism as part of a backward-looking reaction *against* the trappings of modernity (see pp. 98–139). The confusion can be cleared up by noting that what Lears has in mind is chiefly the knightly image of military manhood made popular in the novels of Sir Walter Scott. The ideal of "generalship" which I explore below is unmistakably forward-looking.   [45] Mrozek, "The habit of victory," 226.

[46] Rotundo, *American manhood*, 238; Lutz, *American nervousness*, 93.

mental program became an ideal of generalship. Although this ideal acknowledged the value of organization and division of labor, the assumption was that
organizations could not possibly "run themselves." The emphasis was on
command from above, and the burden of responsibility for success or failure
rested squarely with the man at the top. A passage about leadership from
Walker's biography of General Hancock is useful as a window on Walker's
understanding of his own work:

[I]n many wars the successful leader whom fame thus selects for immortality, actually
did, by his genius, bring into existence all that was above commonplace – was, in effect,
his whole army, in all that compelled victory. He had, indeed, capable and efficient lieu-
tenants to execute his plans . . . regiments, brigades, and divisions were officered by men
who in many actions received deserved praise; while the rank and file were in their place
brave, loyal, and enduring. Yet it still remains true that the General was the army, and
the whole of it, in this sense: first, that had the army, good as it was, been given into
the hands of a soldier less masterful, it would, in the situation existing, have been
beaten; secondly, that had the commander been given an army far less fortunately composed and officered, he would, before the end, have shaped and tempered it until fit for
victory.[47]

In line with this conception, Walker practiced in all his public roles what a later
age would term "micro-management" (with all the advantages and disadvantages that implies), because he assumed it was the only way to proceed.
Combining impartiality with (at first) tireless organizational energy, Walker
lived out a military model of governmental manhood that represents a significant transitional moment between the mid-nineteenth-century "man of
affairs" and the Progressive Era "social scientist/expert" (see Chapter 3 above).
The failure of his governmental generalship coincided with the end of what I
have called "the" (singular) governmental subject; shortly after Walker's
death, programs of national governmentality would cease to be understood as
dependent upon the ability of single individuals to master all their details. In
the remainder of this chapter, I will give an account of the main features of
Walker's generalship, moving from his self-effacing impartiality to his self-
exhaustion and self-extinction. Finally, I will dwell at more length on the wider
meaning of his death.

As I suggested in Chapter 3, the character trait of impartiality can be seen
as a stage preceding the (in principle, total) effacement of individuality personified in the figure of the scientist. Impartiality fits nicely among the virtues
animating the exemplary man, since it could be understood as a special form
of self-sacrifice which nevertheless enhances one's reputation. I have already
touched (in Chapter 2) on the importance of Walker's criticism of the 1870
Census results in establishing his impartiality. One further anecdote should
illustrate the central role played by impartiality in his self-understanding. On

---

[47] Walker, *General Hancock*, 1–2.

the occasion of Walker's death in 1897, one J. J. Spencer published an appreciation in the *Review of reviews* which included a stylized recollection of the pattern of Walker's encounters with the temptations of patronage while overseeing the 1870 Census. Walker was repeatedly approached with the suggestion that his power to appoint clerks would allow him to have "whatever he desire[d]." The ensuing conversations, according to Spencer, followed a set format: "'I have no desires', said General Walker. 'But General', said his coadjutor, 'do you not see that we can push forward our friends and relatives into good places?' 'I have no friends' was the characteristic reply."[48] As Munroe notes in the introduction to his biography, Walker's self-effacement extended also to posterity; unlike many other prominent men, he destroyed all personal correspondence.[49]

Such disinterestedness was for Walker an individual character trait, as he remarked in a public debate with Nathaniel Shaler over the value of technical schools. To Shaler's charge that schools such as MIT typically had too unbalanced a curriculum to foster scholarly habits of disinterested inquiry, Walker replied that such traits depended not on the environment but on the man.[50] Walker repeatedly claimed that the first principle of the scientific education offered at MIT was "a belief in the essential manliness of men." Even in engineering (where the issue of disinterested work could have life-or-death consequences), the main question for Walker was still "how much of a man one is."[51] This was not yet the modern image of the scientist.

A self-affirming self-denial also formed the core of his military work ethic, and led eventually to his self-destruction. He took up whatever work was at hand, by all accounts everyone else's as well as his own. And, like the hypothetical general he describes in the passage quoted above, he made sure that the undertaking (whatever it happened to be) was a success. Again, Munroe provides a wealth of evidence from which I will lift a small sample. In one of the most interesting passages relevant to Walker's work ethic, Munroe places this ethic in a clearly sexualized, neurasthenic context. The immediate subject is a set of lectures Walker wrote for a course he gave on money and currency issues at Johns Hopkins in 1877, but Munroe clearly intended it to stand for many other episodes as well: "after such an orgy of work – for no other term expresses the zeal with which Walker threw himself into an undertaking like the Johns Hopkins lectures – he usually had to give himself a more or less enforced period of rest."[52] It is not so surprising to find Walker expending his nerve force in vast quantities, but one feature of Munroe's formulation stands out. Specifically, it is not an accident that Walker had to "enforce" his periods of rest. His military manhood ideal did not permit easy withdrawal from toil into relaxation. Implicitly, to fall back on something like neurasthenic illness

[48] Quoted in Munroe, *A life of Francis Amasa Walker*, 112.   [49] *Ibid.*, vi.
[50] *Ibid.*, 394–395.   [51] *Ibid.*, 399, 397.   [52] *Ibid.*, 184.

was to admit weakness to an unacceptable degree.[53] If Munroe had to admit that Walker rested, it was imperative that he first emphasize the amazing efforts which demanded rest, and then that rest be portrayed as a demand, a forcing. Thus Walker was implicitly "defeated" by the need to rest; he submitted to it, though unwillingly.

The one instance in which Walker admitted to something like a breakdown occurred during his captivity in the midst of the Civil War. After having been marched to the point of exhaustion, having escaped, swum across a river and then been recaptured, and then having been marched around again, he suffered "a period of nervous horror such as I had never before and have never since experienced, the memories of which have always made it perfectly clear how one can be driven on, unwilling and vainly resisting, to suicide."[54] This must have been a brutal experience, and it was surely not his manhood ideal alone that led him to ponder taking his own life. But I would suggest that the difficulty Walker had in imagining how one would continue to live in a state of nervous horror is related to his larger commitment to a military manhood ideal. The options for a soldier are working and fighting, or death, but not rest or withdrawal.

In his civilian executive capacities as well, Walker was no stranger to situations that taxed his nerve force. But his "reserves" held out for a long time, allowing him to outlast lesser men. The massive census undertakings and, later, the demands of the presidency of MIT, in addition to the continuous and ever expanding volume of formal and informal consulting he did on various governmental subjects, characteristically revolved around managing people, information and a high volume of official communications. It could be a lethal business, as the following passage from one of his magazine articles on the census attests. Many of the themes I have been discussing come together in it, so it is worth quoting at length:

I have said that the necessary agencies for taking this great decennial inventory, which now embraces population, wealth, taxation, industry in all its forms, transportation, education, physical and mental infirmity, pauperism, and crime, have been freely provided by Congress. The only limit now to the usefulness of this great work is found in the limited ability of any one man to grasp so many subjects at once; to make fitting preparations for a canvass of a nation of such territory and population as ours; to build in a few months, from the ground upward, the entire machinery of enumeration; to raise, organize, officer, equip, and instruct an army of fifty or sixty thousand men for this service; to set them at work on the 1st of June, all over the country, from Maine westward to Oregon and southward to Florida and Texas; and thereafter to keep them

---

[53] See McLaren, *The trials of masculinity*, 156, and Rotundo, *American manhood*, 191 on the social Darwinist interpretation of neurasthenia. Although Walker rejected social Darwinism as an economic principle (see Chapter 6), he held himself personally to an ethic of endurance that had a lot in common with it.

[54] Quoted in Munroe, *A life of Francis Amasa Walker*, 93.

at work, vigorously, zealously, unfailingly, to the full completion of this mighty task. The limits spoken of are not theoretical merely. It is a question if those limits – whether as to brain power or as to will power – have not already been reached or overpassed . . . Aside from the question of the Superintendant's intellectual ability to comprehend his work in all its parts, and to make provision for every foreseen occasion and for every sudden exigency of the enumeration, the strain upon the nerve and the vital force of whomsoever is in charge of the census is something appalling. My successor in the Tenth Census, Col. Charles W. Seaton, was literally killed by the work, and three successive chief clerks of that census died in office. The present Superintendant of the Eleventh Census, Mr. Porter, was driven away to Europe by his physician last summer, while the work was at its height, to save his life. Taking a census of the United States under the present system [i.e., without a permanent Census Bureau], and upon the existing scale, is like fighting a battle every day of the week and every week for several months.[55]

As this passage strongly suggests, one of the ways in which Walker made the superintendancy so strenuous for himself (and for those who took his approach as a model for their own) was through his insistence on personally cross-checking many of the statistics being compiled. He applied this micromanagement style to the 1874 *Statistical Atlas* as well, expending great energy in checking the maps for accuracy.

At MIT as well, Walker persevered where others were defeated. The two men who preceded him in the presidency, William Barton Rodgers and John D. Runkle, both bowed out after a few short years at the helm, incapacitated by the strain.[56] After more than a decade at MIT, the (clearly reluctant) Walker "was persuaded to employ" a secretary. For some years before this time, the institute had employed a single secretary for the faculty as a whole, but he was a man, in fact, none other than the James Phinney Munroe who would later write Walker's biography. Walker had already had ample experience of female clerks while at the Census Office. The 1870s saw the first major movement of women into federal offices, with 1,773 (out of a total of 7,866 clerks) working in the Executive Departments in Washington by 1880.[57] But during Walker's years at MIT female clerical workers were still relatively scarce in educational institutions. For all the obvious benefits of additional staff, women would still have constituted quite a cultural shock to men who were accustomed to thinking of the ivied halls as a male preserve.[58] Walker wrote out most of his own correspondence until the early 1890s. Especially in view of the great diversity of his commitments (see Chapter 3), the time burden must have become immense. Even with Miss Holt on board, the pressure on Walker's time was unrelenting, as the following glimpse of the general as he greeted visitors

[55] F. Walker, "The United States Census," *Forum* 11 (1891), 258–267, reprinted in F. Walker, *DES*, vol. I, D. R. Dewey (ed.) (New York, 1899), 100–101.

[56] Munroe, *A life of Francis Amasa Walker*, 214.

[57] C. S. Aron, *Ladies and gentlemen of the civil service: Middle-class workers in Victorian America* (New York, 1987), 5.   [58] Rotundo, *American manhood*, 250.

taking advantage of his open-door policy makes clear: "to each [caller] . . . he would give his seemingly undivided attention – although he might at the same time be writing a letter and consulting, *sotto voce*, with some fellow officer of the Institute – and he would make that caller feel that the particular matter which interested him was the one thing in the world which most closely concerned Professor Walker."[59]

By his later years at the institute, Munroe reports, Walker's various duties and commitments had come to fill up not merely the workday proper, but most of the rest of his waking hours as well: "a great part of his writing, of his preparation for public addresses, of his drawing up of reports for civic and other committees, etc., had to be done at night, in his home. Often, he would accomplish what many men would regard as a full day's work after his return from some banquet, or other public function."[60] Here again, the shadow of his wife looms large, but her exertions go unrecorded. Inevitably, "the sum total of all these varied demands upon his time and vitality was more than any man could bear . . . when urged not to work so hard and for so many causes, he would say, quite simply and convincingly: 'but the work has to be done'."[61] Francis Walker believed that death should take people while they still had "intellect, courage and cheerfulness." He wrote to an inquirer in December of 1896 that "the duality of mind and matter is to me simply unthinkable. I only know force, in its infinite variety of manifestations, from the thought of the poet to the energy of the coal in its deep bed beneath the mountains." Eighteen days after this letter, on January 5, 1897, his force spent at the young age of 57, Walker died of apoplexy.[62]

The most telling memorial to Walker was entered by the journal *Review of reviews* in the February 1897 issue, in the form of a "plea for the protection of useful men." The editor takes a solemn but scolding tone, asserting that "so valuable a piece of public property as such a man ought not to be worried and badgered to death by petty demands on his time and strength."[63]

Only two weeks before the sad news of General Walker's death, the editor of this review received from him the following letter, – a remarkable letter in any case, and well-nigh startling in view of the event that was soon to follow: . . .

"Dear Mr. Shaw:

. . . I should be glad sometime to write an article – but probably never shall – having for its title, 'Killing a man,' in which I should try to set forth the manners and ways in which decent and well-meaning people combine and conspire to knock down and trample on every man in the community who is fit to render any public service. I should try to show what an utter lack of conscience there is in this matter, so that men who

---

[59] Munroe, *A life of Francis Amasa Walker*, 371.   [60] *Ibid.*, 369.   [61] *Ibid.*, 260–261.
[62] *Ibid.*, 403–405, 402, 401.
[63] Anon., "A plea for the protection of useful men," *Review of Reviews*, Feb. 1897 (n.p.), in Francis A. Walker Papers, Massachusetts Institute of Technology Archives and Special Collections, MC 298, Box 5.

would not on any account commit a petty larceny, will set upon a man whom they per- fectly well know to be badly overworked, and knock out whatever little breath may be left in his poor body; how they get 'between him and his hole,' cutting off his possible retreat by every sort of social entanglement; how they make last year's declination a reason for this year's acceptance; how they surround the poor victim on every side until he is fain to surrender and give up the last chance he has of getting a little rest or a little pleasure during the next two weeks, all for the purpose of delivering an address for some infernal society, which, perhaps, ought never to have existed, or at any rate, has long survived any excuse for its being."[64]

Since my purpose is not to write a biography of Walker, but rather to employ him as a representative of important changes that occurred during his lifetime, I would like to end this chapter not with Walker's end, but instead by drawing a parallel between his death and the death three years earlier of William F. Poole. Poole, like Walker, was one of the first generation of American men who would devote large parts of their careers to the systematic organization of information at a national scale (see the discussion of the rise of indexes and catalogue systems in Chapter 3). Poole's undertaking was not strictly speaking a matter of governmentality (at least as I have been using the term), but a brief account of his demise should reinforce the impression that it was especially the information gathering aspects of governmentality which posed a challenge to late nineteenth-century masculinity.

Poole, born in 1821, eventually became the head librarian at Chicago's Newberry Library, and his life's great labor was the compilation of *Poole's index to periodical literature*, the famous third edition of which appeared in 1882.[65] This was one of the first attempts exhaustively to catalogue all of the periodical articles that had appeared in "significant" American magazines (and some of the better known British journals as well) from the founding of the republic until the 1880s. Subsequent supplements updated the index at roughly five-year intervals. While he lived, Poole himself made the decisions about what constituted "significance" and he left no precise record of his cri- teria, but the magazines canvassed in his index numbered in the hundreds. Poole was active in the public library movement, and in organizing national and international library conferences. The peak of his organizational activity was apparently the Columbia Exposition of 1893 (though his chronicler does not specify the exact nature of this activity).[66]

What concerns me here is the way Poole approached the organizational problems posed by his index, and the way his disciple Fletcher (much like Munroe with Walker) related his working life to his manhood and to his death. Poole's character was not quite as "hard" and martial as Walker's, but in many respects it was similar. He was

---

[64] *Ibid.*   [65] W. F. Poole (comp.), *Poole's index to periodical literature*, 3rd ed. (Boston, 1882).
[66] W. I. Fletcher, "Obituary for William F. Poole," in W. F. Poole and W. I. Fletcher (comps.), *Poole's index to periodical literature: The third supplement* (Boston, New York, 1897), iii–iv.

a fine example of the *mens sano in corpore sano,* enjoying perfect health all his life. His commanding presence and benign expression were truly representative of the strength and sweetness of his mental and moral character . . . In his life as in his writings, he showed a native instinct for that which is honorable and of good report, and a corresponding dislike of all sham and pretense. No qualities were dearer to him than candor and sincerity, and it was the lack of these elements in the historical writings he reviewed that provoked his sharpest censure.[67]

This healthy, honest man was confronted (like Walker) with a monumental task: to collect and organize citations for all articles in hundreds of magazines over the course of the nineteenth century. Where Walker's "army" had consisted of census enumerators, Poole enlisted the help of fifty-one cooperating librarians around the country. In subsequent supplements, the number of collaborating libraries would rise into the sixties, and editors of some of the magazines would provide more and more of the indexing help. The difference between Walker's approach to censuses and Poole's to his periodical index was that Poole relied more comfortably on delegation of tasks.

The cooperative feature of this work will attract the attention of persons interested in this special phase of social science . . . As the person who has had the sole management of all the details of this enterprise, I desire to put on record this testimony in behalf of the cooperative principle. It is simple, effective, and attended with no embarrassment or difficulties of any kind.[68]

Perhaps because his troops were skilled librarians rather than the untutored citizens who acted as census enumerators, Poole could avoid the most taxing features of generalship.

Nevertheless, it was a difficult job, as Poole and Fletcher made clear in the preface to the first supplement:

those who may have been impatient at the delay [in bringing the first supplement out] will hardly expect an apology when they look at the list of contributors and their work, and see that a large share of the indexing itself has fallen upon the editors [Poole and Fletcher]; and are further informed that this work of indexing, with that of arrangement, revision, and proof-reading, has been done by busy men during hours which other workers devote to rest and recreation.[69]

This work, in combination with his organizational labors connected with the 1893 Columbia Exposition and the removal of the Newberry Library to a new location "doubtless undermined his system, and he died on March 1, 1894, after a brief and apparently trifling illness." Before being carried away, Poole had withdrawn from the indexing work out of exhaustion.[70]

It is interesting that in the accompanying preface by Fletcher, the "trifling

---

[67] *Ibid.*, iv.   [68] W. F. Poole, "Preface," in *Poole's index to periodical literature*, 3rd ed., xi.
[69] W. F. Poole and W. I. Fletcher, "Preface," in *Poole's index to periodical literature: The first supplement* (Boston, New York, 1888), iii.   [70] Fletcher, "Obituary for William F. Poole," iv.

illness," the withdrawal from indexing, and all other warning signs disappear, allowing Poole's death to acquire the same sort of manly suddenness which Munroe ascribed to Walker's: "In a little more than a year after the date of [the previous] preface the life that had been so strong and vigorous that one hardly associated age with it, even at seventy, was suddenly terminated in the midst of the activities of a busy and useful career."[71] Poole, experimenting with a mode of large-scale organization less dependent on "the man at the top," nevertheless found his involvements too taxing, and like Walker, ended up a sort of epistemological martyr to the emergence of modern America.

I have included the story of Poole in order to bolster my claim that Walker's failure was symptomatic of a larger pattern in which inherited models of manhood were being stretched to breaking point by the informational and organizational requirements of attempts to master various aspects of the American social body. These requirements constituted a special version of the crisis of manhood outlined in the first part of this chapter, and added an additional strain to the already considerable duress under which even those (such as Walker) who had "proven" their manhood by normal standards were forced to labor. Because Walker was so deeply immersed in the vigorous lifestyle required of someone who was understood (and who understood himself) as a minor icon in the wider "cult of true manhood," and because his "governmental subjectivity" added an ultimately insupportable strain to these demands, we can easily understand how the crisis of manhood came to form such a central thread in his program of governmentality. This program is the subject of Part II.

---

[71] W. I. Fletcher, "Preface," in *Poole's index to periodical literature: The second supplement* (Boston, New York, 1893), iii.

Part II

As Part II will make plain, Francis Walker displayed a highly sensitive awareness of spatial issues. In keeping with my commitment to portray Walker's career in a "structural" rather than a "biographical" way, I would like to introduce this part of the book by briefly tracing his spatial awareness to the series of subject-positions he held during the early phases of his public career. The particular ways in which he refined and used this awareness will be described in greater detail in subsequent chapters. The terms I have italicized here foreshadow more precise analytical concepts developed in Chapter 5.

Structurally speaking, Walker's unusual geographical awareness is most obviously related to his experiences in the Civil War and with the census. It requires no great leap of conjecture to imagine that his presence at numerous high-level staff meetings thoroughly inculcated a sense of territory as an *abstract* object of strategies of control, and a sense of groups of people having their proper places in the landscape. His position as assistant adjutant general entailed keeping administrative track of the personnel and equipment deployed across the landscape, and would therefore have given Walker a sense of the importance of *assortment*. One could easily imagine that Walker's unusually strong predilection for maps and other graphic media originated in the crucial, probably at times desperate, importance of having an accurate synoptic view of events on a battlefield. Not only was accurate information essential, but smoothly functioning centralized coordination would also have been an obvious key to military success. Since part of Walker's job was to disseminate orders and receive and collate reports from the field, Walker also had ample reason to develop an interest in the quality of distant acts of observation and the importance of efficient *compilation* at the command center. It is quite likely also that, especially when operating in Confederate territory, Walker acquired a sharper appreciation of the importance of citizen cooperation in effective territorial control.

Walker's fellow "best men" among the Union officer corps (Charles Francis Adams and Oliver Wendell Holmes, Jr., for example) would have shared many of these experiences, but were not in positions so centrally concerned with the logistics of gathering and using precise information. The emphasis his particular duties gave to Walker's wartime experience was then, of course, strongly reinforced when he became superintendent of the census. As noted in the discussion in Chapter 2 on "grids of specification," the mere fact that census information was tied to specific locations would have reinforced the assumption that people have their natural places in the territory.

While his census duties were very different from his wartime duties in the sense that they involved mastering territory for benign rather than hostile purposes, Walker understood that the distinction was not as clear and simple as it seemed. In retrospect, it is difficult to imagine a sequence of professions more perfectly suited to produce a refined sense of the relationships between space and power. It is little wonder that Walker took so quickly to his position as head of the Bureau of Indian Affairs in 1871. The links between military coordination, census-taking and administration of the reservation system would have been patently obvious.

I believe it is possible to go further, and argue that Walker's display of what David Sibley terms an "exclusionary psychology" is also traceable to this series of public positions. Sibley's interesting study *Geographies of exclusion* offers a number of ideas about the connections between psychological responses to unfamiliar "others" in modern Western culture and spatial policies and practices of exclusion that have been a notable feature of social policy in the West. One of his main points is that "[t]he construction of community and the bounding of social groups are a part of the same problem as the separation of self and other."[1] Many of the psychological patterns Sibley traces seem to resonate closely with Walker's motivations, and thus may help explain the strength of his impulse to make restriction and fixation a part of his regulatory programme. While it would be impossible to separate out the relative influence on Walker's policies of more or less conscious, explicit beliefs versus subconscious, psychological reflexes, it is quite likely that the two often pushed him in the same direction, and thus reinforced each other. My discussion of Walker's exclusionary psychology will be brief but (I hope) suggestive.

To start with, the feelings of insecurity which Sibley identifies as an important source of exclusionary practices undoubtedly animated Walker and the rest of the native, white, male northeastern cultural elite during the late nineteenth century (see Chapter 3). Sibley observes that "[f]eelings of insecurity about territory, status and power where material rewards are unevenly distributed and continually shifting over space encourage boundary erection and the rejection of threatening difference. The nature of that difference varies, but the imagery employed in the construction of geographies of exclusion is

[1] D. Sibley, *Geographies of exclusion: Society and difference in the West* (New York, 1995), 45.

remarkably consistent."[2] Among the standard patterns in the Western European tradition has been the dual association of the purity of the home community or nation with whiteness, and on the other hand, of the impure, foreign "other" with dirt or non-white colors.[3] This in turn is an example of what Sibley calls the "epidermalization" of defiled, stereotyped others, the assumption that essential inferiority or uncleanliness must have an outward manifestation.[4] It is not too much of a stretch to see Walker's central imagery of "unhung gates," dilapidated shutters and "green pools" in immigrant front yards as an example of such epidermalization (see Chapter 6). Epidermalization would fit nicely with the strong visualism that played so prominent a role in his conception of census information (see Chapter 5), in his model of the ideal modern citizen-worker and citizen-scientist (see Chapter 7) and in his theory of the threat to native white birth rates (see Chapter 6).

Walker certainly displayed the trait, which Adorno had ascribed to the "authoritarian personality," of a dislike for ambiguity, especially when it came to matters of race. His relatively sophisticated empiricism allowed Walker to be far more sensitive than many of his colleagues to complexity, but complexity and ambiguity are different. One of the things that seems to have attracted Walker to statistics was their ability to eliminate ambiguity while making ample room for complexity. Hostility to ambiguity may in part account for his efforts to establish "foreignness" as a real and lasting social category, for example in his insistence on adding a question to the 1870 Census on foreign parentage as well as foreign birth. After all, what could be more troubling than the fact that all white "native" inhabitants of the United States had at one time been "foreigners" and immigrants? It must have struck him as imperative to find some way of distinguishing native from foreigner in a way that would lend scientific authority to his conviction that the difference was real and important. Similarly, ambiguous *spaces* are likely to be troubling to someone seeking to protect purity. Sibley notes that "entrance zones" such as the doorways of households, can be the focus of much anxiety.[5] The analogue at the national scale for Walker was the Castle Garden immigrant arrival terminal in New York City, and his concentrated animus against the "beaten men from beaten races" pouring through this national front door may well have ratcheted up the vitriol with which he often spoke and wrote about New York City (see Chapter 5). Yet another spatial form of ambiguity identified as significant by Sibley also troubled Walker: the condition in which groups stereotyped as belonging in specific places turn up "out of place." Again, partly because of the structure of census information, Walker had very strong ideas about who belonged where. While he saw earlier generations of European immigrants to

---

[2] *Ibid.*, 69.   [3] *Ibid.*, 24.   [4] *Ibid.*, 18.   [5] *Ibid.*, 33.

the United States as self-selected by their unusual ambition and vigor, he felt that the peasants pouring in from Southern and Eastern Europe starting in the 1880s were being unnaturally introduced into the American social body, especially when recruited by steamship companies in Europe (see Chapter 6). These people were quintessentially "out of place," and aroused great exclusionary impulses in Walker (and others).

Finally, Sibley identifies the rhetoric of disease and contagion as a common code for the threat of racial mixing, which was certainly widely feared by people of Walker's ilk in the late nineteenth century.[6] Walker did not often have much explicitly to say about this fear, but his increasingly essentialist racism, centered as it was on "blood," strongly implies that it was among his worries. The blood he feared was already at large within the national territory; thus, he must have been tempted to advocate segregation of some sort for immigrants. Together with the other elements of his exclusionary psychology, as well as his highly developed understanding of who belonged where, his fear of racial mixing heavily predisposed him toward segregative policies. Unable to justify internal segregation of immigrants, he had to settle for *international* segregation.

It may be troubling to find this part of the study introduced by a discussion of the psychology of an individual subject, given my heavy reliance on Foucault. Both the categories "psychology" and "individual subjectivity" are labels for derivative social constructions, each helping to mask the constructed nature of the other. As if this weren't enough to violate the strictures of Foucauldian research, I also propose to analyze the workings of governmentality through the work of a single individual. Anticipating and responding to these objections is worth a brief pause. My justification for such an unorthodox turn of argument harks back to the critical discussion with which I opened Part I. Once again, an important goal of this study is to illustrate the potential for articulations between Foucauldian social constructionist analysis and more traditionally "realist" approaches.

At the core of the possibility for rapprochement lies the recognition that social constructionist analysis of the derivativeness and contingency of seemingly natural phenomena does not preclude the recognition that these phenomena have causal efficacy, internal dynamics, in short, reality. This is not a new insight, but it is less often stressed than one might expect, perhaps because of the polarizing nature of academic debate (for example, the indiscriminate use of "slippery slope" arguments to maintain differences of opinion). In the context of my argument, this simple insight allows me to follow a careful explanation of Walker's constructedness (Part I) with a detailed account of his thoughts and actions in their historical effectivity (Part II). Because constant reminders of Walker's derivativeness throughout Part II would expand this

[6] *Ibid.*, 25.

study substantially, it will be necessary for the reader to recall periodically that the various discourses I have assembled under the roof of "Francis Walker" are, in many other instances, dispersed among different people and institutional sites. Again, it is only because Walker was such an uncommonly "extensive" subject that I have the luxury of personifying the governmental logic that is my chief focus.

Part II of this study arranges Walker's work according to the "cycle of social control" identified in Chapter 1 as the root structure of "power/knowledge" (whether of the disciplinary or the governmental sort). In Chapter 5, I lay out the spatial politics of the first "moment" of the cycle of social control, governmental *observation*, as these politics were manifested in Francis Walker's work on the Ninth and Tenth US Censuses. In Chapter 6, Walker's political economy will provide the interpretive framework through which to view his *normalizing judgments* (the second "moment" of the cycle of social control) about whether and how the national social body known (and formed) through the census needed to be regulated. Chapter 7, which rounds out Part II, will be devoted to showing how Walker's position on immigration restriction represents a logical resolution of the tensions within his spatial politics, and reconciles his ideas on race and gender as well.

5

The spatial politics of governmental knowledge

To say that Francis Walker had a "program for governmentality" is of course to construct his thought using interpretive tools only available a century after his death. My purpose, again, is not to write a biography (intellectual or otherwise), nor is it to represent Walker's thought as he might have understood it. I impose a structure on his writing and career in the service of a history of the present. In this and Chapters 6 and 7, I will arrange Walker's thought and activities around a general logic of governmental power, rather than as a temporal progression in his ideas or institutional subject positions.

Here, I will frame the analysis with a brief discussion of the parallels between colonial regimes of knowledge and governmental observation "at home" in the metropolitan world, in order to bring out as clearly as possible the political character of large-scale projects of knowledge gathering. By using the adjective "political" and the noun "politics," I mean to signal the contestable character of such projects, and the futility of attempting to naturalize them as "merely rational." Only by getting beyond scientist assumptions is it possible fully to understand why Walker and others had to (and still must) struggle for their systems of knowledge. The next two sections will form the heart of the chapter. In the first of these, I explore the spatial politics of field observation, which revolve around what I term abstraction and assortment (by "spatial" politics, I mean contested attempts to improve knowledge-gathering by reorganizing its spatial aspects). Abstraction and assortment will be seen to work in tension with one another, to embody a fundamental contradiction between mobility and fixity. The second main section is devoted to the analysis of what Latour calls "centers of calculation." Walker's experience as superintendent of the census offers ample material with which to illustrate the political character of institutional centralization and compilation. These issues lead, finally, into a reading of Walker's 1874 *Statistical Atlas of the United States* as an embodiment of his observational politics and a powerful construction of the American social body.

Colonialism and the politics of governmental observation

[M]odern statecraft is largely a project of internal colonization, often glossed, as it is in imperial rhetoric, as a "civilizing mission." The builders of the modern nation-state do not merely describe, observe, and map; they strive to shape a people and a landscape that will fit their techniques of observation.[1]

The connection drawn here between "home" statecraft and colonial administration has attracted the notice of a number of scholars.[2] Colonial regimes have often served as "theatre[s] for state experimentation" within which Western powers have attempted to work out efficient methods of social control.[3] Imperial state officials have frequently assumed greater freedom to experiment abroad than at home, justifying this freedom on the basis of theories of European "racial" superiority. They have also often benefited from the relative weakness (at a national scale) of the kind of Western-style civil society among colonized peoples that could most effectively appeal to the political conscience of metropolitan electorates.[4] Colonial states have often been "more bureaucratized and less tolerant of popular resistance" than metropolitan governments.[5] Especially where imperial powers have not been dependent on the pre-existing indigenous economies of colonized peoples (as, for example, with Native Americans), there has been little incentive for colonizing states to pull back from pursuing total control of subject populations. Although such comprehensive subjection has never been achieved in practice, its pursuit has tended in some cases to give a full-blown panoptic flavor to colonial administration. In principle, the kind of governmental thinking characterized by a more careful and solicitous approach to social control has been less prominent in imperial regimes.

But the distinction is actually not so clear-cut. There are many ways in which colonial rule can acquire more "governmental" characteristics, and conversely, there are important senses in which even quite "benign" governmentality retains a basically "colonial" structure. The most basic element of this structure is *rule from a distance*. Both national-scale governmentality and colonial rule involve the making of decisions in distant centers which affect local conditions of life for subjects or citizens. In precolonial or pregovernmental times, the activities and decisions of different people in different places

---

[1] J. Scott, *Seeing like a state: How certain schemes to improve the human condition have failed* (New Haven, CT, 1998), 82.

[2] M. Edney, *Mapping an Empire: The geographical construction of British India, 1765–1843* (Chicago, 1997); D. Gregory, *Geographical imaginations* (New York, 1994); B. Cohn, *Colonialism and its forms of knowledge: The British in India* (Princeton, NJ, 1996).

[3] N. Dirks, "Foreword," in Cohn, *Colonialism and its forms of knowledge*, xi; Paul Rabinow's *French modern: Norms and forms of the social environment* (Chicago, 1989) is especially thorough in connecting colonial and metropolitan administrative programs through the small community of eminent French planners who moved back and forth between the two contexts.

[4] Scott, *Seeing like a state*, 5, 49.   [5] *Ibid.*, 69.

within a particular region would have been integrated "horizontally," that is, through arrangements undertaken and regulated by people living in the linked places. The advent of distant centers of government brings with it a tendential shift to more "vertically" integrated relationships between places. In other words, more decisions about how and to what extent particular activities and people in particular places should be connected or related to other people and places come to be made elsewhere, on the basis of criteria other than local knowledge and local calculations of interest or necessity.

In making this argument, I am playing devil's advocate to some extent, in order to highlight the unilateral, "imposed" aspects of governmentality. Clearly, the at least nominally democratic contexts within which modern governmental systems are nested make some difference in the degree of onerousness of rule from a distance, and subjection to such rule certainly has tended to come with far more substantial rewards for citizens of sovereign nations than for colonial subjects. In addition, the impact of racial ideologies in subjecting the colonized to serious indignities never suffered by most metropolitan citizens should not be so lightly passed over. Finally, it must be admitted that many forms of governmentality organized on a national scale (for example, social security systems or veterans' pensions), far from being imposed upon populations, have been actively sought by them. Nevertheless, despite the relatively benign, positive character of governmentality, as I use it here it remains a form of control, and there is much in it to trouble those subject to it.

Colonial and home rule are similar especially in the epistemological register, as James C. Scott shows so nicely.[6] In either setting, state power requires at least the mapping and bounding of territory, the identification and registration of people living within it, and an inventory of resources. In both settings, these requirements for knowledge are acted upon unilaterally, without seeking the consent of the people. Furthermore, in the metropole as well as in the colonies, knowledge gathering has always been an attempt to (re)order as well as to know social reality. As Scott puts it,

Processes as disparate as the creation of permanent last names, the standardization of weights and measures, the establishment of cadastral surveys and population registers, the invention of freehold tenure, the standardization of language and legal discourse, the design of cities, and the organization of transportation [are] comprehensible as attempts at legibility and simplification . . . Thus a state cadastral map created to designate taxable property-holders does not merely describe a system of land tenure; it creates such a system through its ability to give its categories the force of law.[7]

Legibility, as well as more obviously coercive conditions, must be imposed if social control is to be at all effective.

---

[6] *Ibid.*, chs. 1 and 2.   [7] *Ibid.*, 2–3.

The idea that governmental knowledge is coercive may seem to run counter to the "liberal" logic that distinguishes governmentality from other forms of modern power. To simplify the discussion in Chapter 1 above, this logic states that because a social body has its own laws, and because those seeking to govern it can only ever have incomplete knowledge of it, regulation should be undertaken only when necessary, and only with great caution. But the "incompleteness" of knowledge had a very specific meaning in late nineteenth-century political economy. It referred to the inability to predict specific outcomes of the operation of social and economic laws. The paradigm case of such a law was the "law of competition," and it was generally agreed among mainstream political economists that the winners and losers in business competition could not be predicted beforehand. But predictive knowledge of specific outcomes was not the only kind of knowledge that could be involved in governing a nation; *a posteriori* knowledge of established facts could also be immensely useful, not least in allowing predictive knowledge of general patterns. The "statistical laws" with which Walker and many contemporaries were so fascinated illustrate the difference: while it would be impossible to try to predict beforehand exactly who would die in Baltimore in 1889, it would be possible to predict roughly how many would meet their end. And it would be possible and useful to record precisely who had died after the fact, along with other information about the deceased which might allow more precise general predictions. Census taking is in this sense the quintessential *a posteriori* knowledge gathering exercise, and the liberal principle of limited knowledge does not apply to it. Thus, while governmentality often differs from colonial or other aggressive forms of social control at the stage of regulatory decisions (see Chapter 7), there is no such clear difference at the stage of knowledge-gathering.

Bernard Cohn identifies six broad forms of colonial knowledge, or "investigative modalities," which together help to situate census taking among other technologies or "layers" of object formation: the historiographic, the observational/travel, the survey, the enumerative, the museological and the surveillance modalities. Without going into the details of these modes, it is worth pausing to note that they have many parallels in the set of cultural forms identified by Benedict Anderson as more generic components of nation building. As Anderson notes, these forms, despite the specificity of their historical and geographical origins, eventually acquired a "modularity" allowing them to be transplanted "to a great variety of social terrains, to merge and be merged with a correspondingly wide variety of political and ideological constellations."[8] Anderson and Thongchai Winichakul do a great deal to demonstrate the inadequacy of a simple distinction between Western "home societies" and colonized peoples by calling attention to the ways in which postcolonial (or in the

[8] B. Anderson, *Imagined communities*, red. ed. (New York, 1991), 4.

case of Thailand, never-colonized) indigenous elites have made use of the techniques of colonial rule to consolidate their own internal hold upon national power.[9] This point reinforces my claim that there is an important political dimension to the techniques of governmentality which has little to do with the question of how "benign" it is in a comparative sense. Five of Cohn's six investigative modalities (all except the historiographical) clearly involve a spatial politics of one sort or another, but in only three (the survey, enumerative and surveillance modalities) can the spatial aspects of knowledge-gathering be considered the primary locus of political significance.[10] I will concentrate on the enumerative modality in its relation to that of surveillance, but the analysis of spatial politics will be directly relevant to the survey modality as well. Indeed, I will draw heavily on (and in some ways extend) the insights in Matthew Edney's study of the trigonometrical survey of India. Census taking and mapping share much in the way of fundamental political structure.

The spatial politics of the observational field: abstraction

There are a number of ways in which one could organize an analysis of governmental observation, and the interpretive structure chosen here is not inherently superior to all others. I take as my cue the metaphor of vision, with its distinction between the field of vision and the viewer, and ask what conditions have to obtain with respect to both in order for something like a census to be taken. It will quickly become clear that the distinction between the field of observation and its observers is not so cut and dried, but it is quite fruitful as a heuristic device. Many of Francis Walker's ideas about improving the census, his projects, struggles and expostulations, attach quite nicely to one or the other side of it. I argue that the politics of the field of observation necessarily involves *abstraction* and *assortment*. Under the heading of abstraction fall the establishment of grids of reference, physical access to all parts of a territory, and, of most interest here, access to information. I will be concerned in detail only with the last of these issues, since Walker had much less to say about the others. By assortment I mean the conditions which allow the objects

---

[9] *Ibid;* T. Winichakul, *Siam mapped: A history of the geo-body of a nation* (Honolulu, HI, 1994).
[10] Important studies which treat the spatial politics of the historiographic and observational/travel modes include: J. Blaut, *The colonizers' model of the world: Geographical diffusionism and Eurocentric history* (New York, 1993); Gregory, *Geographical imaginations*; E. Said, *Orientalism* (New York, 1979). Matthew Edney's excellent study of the mapping of India (*Mapping an empire*) is probably the most sophisticated full-length work by a geographer on the survey modality; J. Murdoch and N. Ward's paper on the British Agricultural Survey ("Governmentality and territoriality: The statistical manufacture of Britain's 'national farm'," *Political Geography* 16, 4 (1997), 307–324) provides a strong analysis of the survey as a governmental exercise. The enumerative modality is the subject of Bernard Cohn's "The census, social structure and objectification in South Asia," in *An anthropologist among the historians and other essays* (New York, 1987), 224–254.

of census taking to be individually distinguished at fixed locations within the field of observation. The concrete context for Walker's pursuit of abstraction and assortment was his superintendency of the 1870 and 1880 Censuses. The 1870 Census was taken under the 1850 law, and Walker considered it highly inadequate for that reason; the 1880 Census was taken under new legislation essentially written by him. Thus, the more conventional narrative underlying my discussion of abstraction and assortment is the story of Walker's attempts to compensate for defects in the 1870 count and of his improved system for 1880.

As noted above, grids of reference (established by trigonometric, topographical or cadastral surveys, for instance) are not the focus of this study. However, they do form an important prerequisite for census taking, and thus bear schematic treatment.[11] It is impossible to undertake an accurate census unless there is some geographical framework on which to define and precisely locate enumeration districts which exhaust the territory without any overlap. Unlike European national territories, much of the American landscape (rural and, especially in the west, urban) was structured according to reference grids in advance of white settlement.[12] Francis Walker and the Census Office did not need to struggle to establish these grids; other than in a few remote western areas, they could count on a nation substantially mapped in at least a rudimentary topographical sense. The larger point is that the census would have been utterly impossible without the millions of person-hours of surveying and mapping that had already been invested in the landscape.

Nevertheless, these grids were the focus of an interesting politics. In 1870, with only 6,400 "assistant marshals" available nationwide to count the population, enumeration districts could engross as many as 20,000 people, and in the most sparsely settled areas might be as large as or larger than counties. The grand divisions out of which enumeration districts were carved were the federal judicial districts over which the marshals had jurisdiction, and these were of course not designed with a view to census taking. Establishing the boundaries between enumeration districts was often difficult to do with great precision.[13] Walker complained in his 1872 Report to the secretary of the

---

[11] See Edney, *Mapping an empire*. In so far as scholars continue to work out the various components of national-scale governmentality as a geographic enterprise, it will be helpful to think about how, at a macro-scale, these components relate to each other.

[12] See C. White, *A history of the rectangular survey system* (Washington, DC, 1983); J. Corner and A MacLean, *Taking measures across the American landscape* (New Haven, CT, 1996) is a fascinating study of the visible results of this peculiarly American experience with grids of reference.

[13] F. Walker, "Enumeration of the population, 1870–1880," *Publications of the American Statistical Association* 2 (1890), n.p.n, reprinted in Walker, *DES*, vol. II (New York, 1899), p 65; F. Walker, "Interview of the Select Committees of the Senate of the United States and of the House of Representatives to make provisions for taking the Tenth Census," Dec. 17, 1878, *Senate Misc. Documents* 26 (45th C., 3rd S.), in Francis A. Walker Papers, Massachusetts Institute of Technology Archives and Special Collections, MC 298, Box 3, Folder 28, 4–5.

interior that the US marshals who had sole control over the creation of enumeration districts were encouraged to make these districts as large as possible, both so as to minimize the number of enumerators they were forced to supervise, and because a larger district would promise higher pay for the assistant marshals, and thus would be plums worth "purchasing" from the marshals at a higher price.[14] The marshals generally relied on townships and named towns as the basis for enumeration districts, but in some cases, the results were reported in units such as river or creek valleys.[15] In 1880, under a new census law, 31,500 enumerators blanketed the land.[16] The work of the supervisors appointed by Walker, who were charged, among other things, with proposing the boundaries of enumeration districts, was made considerably easier. For one thing, it was far more often possible to assign an enumerator a single township, rather than a group of townships. Second, Walker gave the supervisors more specific instructions on defining the enumeration districts: the boundaries of each district were "to be clearly described by civil divisions, rivers, roads, public surveys, or other easily distinguished lines."[17]

A more subtle spatial politics was enacted wherever grids were present not merely as reference lines on maps but also inscribed on the landscape as street patterns or other easily visible lines. As James Scott notes, "delivering mail, collecting taxes, taking a census, moving supplies and people in and out of the city, putting down a riot or insurrection, digging for pipes and sewer lines, finding a felon or a conscript (provided he is at the address given), and planning public transportation, water supply and trash removal are all made vastly simpler by the logic of the grid."[18] Rule at a distance, whether colonial, national or municipal, is made more efficient if the agents of government can have quick access to subjects without depending on their local knowledge: "for an outsider – or a policeman – finding an address [on a rectangular urban street grid] is a comparatively simple matter; no local guides are required."[19] The 1880 instructions to supervisors on the boundaries of enumeration districts follow this logic. Whereas in 1870 some districts were stream valleys, whose boundaries would have been defined by ridges or highlands of some kind, in 1880 the stream itself would act as the boundary. Although perhaps odd from the perspective of local communities, this method had the advantage of making it far easier to be clear about whether a particular household lay in a particular district. Hypothetically speaking, an assistant marshal in

[14] F. Walker, "Report of the Superintendent of the Census to the Secretary of the Interior," in Office of the Census, *Ninth Census*, vol. I, *The statistics of the population of the United States* (Washington, DC, 1872), xxiv–xxv.

[15] Office of the Census, *Ninth Census: The statistics of the population of the United States*, vol. I (Washington, DC, 1872), 75–296.

[16] F. Walker, "Instructions to enumerators" May 1, 1880, reprinted in C. Wright and W. Hunt, *History and growth of the United States Census*, Senate Document 194 (56th C., 1st S.) US Serial Set 3856 (Washington, DC, 1900), 167.   [17] Wright and Hunt, *History and growth*, 60.

[18] Scott, *Seeing like a state*, 51.   [19] *Ibid.*, 56.

hilly country would often have to rely on the opinions of the inhabitants themselves as to which stream valley they belonged in. In 1880, by contrast, determining the district of a household in a river valley would be a matter simply
of registering which side of the stream it was on.

In short, reference grids afford epistemological control of a territory not
merely in the abstract, symbolic sense of enabling government to "survey the
realm,"[20] but also in the sense of facilitating physical access by the agents of
governmental knowledge. Physical access is another aspect of abstraction
whose thorough treatment is beyond the scope of this study, and which was
not much dwelt upon by Walker, but which nevertheless deserves mention. As
Matthew Edney puts it, "to examine the world requires the examiner to move
purposefully through geographic space and the stationary objects of observation."[21] Even where rectangular grids were absent, a rudimentary road
network of some sort was a great help in providing enumerators with access
to the population. This point is well made by Walker in the midst of an explanation of the importance of competent and literate enumerators. His topic is
the taking of the census in the Reconstruction South, and his larger argument
is aimed at patronage:

[I]t would be a work of the greatest difficulty for a man of more than average intelligence, with an instinct for topography and a fair knowledge of woodcraft, and accustomed to the saddle, to traverse a district containing 400 square miles, in a broken and
wooded country, and not, in spite of the utmost diligence and fidelity, fail to come
upon scores of cabins, hidden away in ravines, or in the depths of forests, often without
so much as a bridle path leading up to the door. It would often be no small task to find
such a cabin, even if you knew it was somewhere in the neighborhood, and were spe-
cially looking for it and it alone. The chance of missing it, when you had no information of its existence, and were only looking around for human abodes in general, would
be very great indeed.[22]

It was perhaps only in the backwoods of Kentucky or Tennessee that such a
comprehensive frustration of official vision would have been encountered, but
the hopelessness of attempting to count people in the absence of a transport
network should be clear.

Physical access was more of a problem in 1870 than in 1880, not so much
because the national infrastructure of roads and rails was steadily growing
denser, faster and more complete in its coverage (railroad track mileage
doubled from 74,096 to 149,214 miles in the twelve years from 1875 to 1887,
for example),[23] but rather because each census enumerator had far more territory to cover in 1870. Since assistant marshals (called enumerators starting

---

[20] See Edney, *Mapping an empire*, Ch. 1 for an elegant account of how British ambitions in India
were constructed as well as reflected in the evolution of maps of South Asia.   [21] *Ibid.*, 53.

[22] Walker, "Enumeration of the population, 1870–1880," 63.

[23] D. Meyer, "The national integration of regional economies, 1860–1920," in R. Mitchell and P.
Groves (eds.), *North America: The historical geography of a changing continent* (Totowa, NJ,
1987), 323.

with the 1880 Census) were locally hired, the issue of physical access was fundamentally local in nature, and the level of national integration of the transport network was chiefly important in getting the completed returns to Washington in a timely fashion.[24] Although Walker complained in his 1870 final report about the slow trickle of returns from around the country, it is unlikely that the national transport network constituted the main obstacle to quick compilation and publication. The key question was how far enumerators had to travel to cover their districts, and how easy or difficult the system of roads, paths, rivers, canals and/or railways made this task.

The final component of abstraction, the final necessary characteristic that renders the field of observation visible, is access to information. It was here that Walker had the most to say. Although the enabling legislation of each census was clear in stating that full, truthful disclosure of information was required of each household (and in theory, enforceable through the threat of a fine), Walker knew that matters were not quite so simple on the ground. Enumerators faced the need to rely on citizens' accounts rather than on the sort of truly unmediated observation they would have had if empowered to line families up in the front yard. And, given the unavoidably intrusive character of census inquiries, distrust was common. In his instructions to assistant marshals in 1870 (and to enumerators in 1880), Walker took great pains to encourage the proper attitude. In 1870, he urged assistant marshals to

make as little show as possible of authority. [Assistant marshals] will approach every individual in a conciliatory manner; respect the prejudices of all; adapt their inquiries to the comprehension of foreigners and persons of limited education, and strive in every way to relieve the performance of their duties from the appearance of obtrusiveness. Anything like an overbearing disposition should be an absolute disqualification for the position.[25]

In the 1880 instructions, he noted that "a rude, peremptory, or overbearing demeanor would not only be a wrong to the families visited, but would work an injury to the Census by rendering the members of those families less disposed to give information with fullness and exactness."[26] According to Walker, Americans were in general favorably disposed toward census takers, since they understood and appreciated the importance of census results more than the citizens of most nations.

It is only where information required by law is refused that the penalties for non-compliance [at that time, a fine of up to $100.00] need be adverted to. The enumerator will then quietly, but firmly, point out the consequences of persistency in refusal. It will be

---

[24] The railroad system was far more important to control of the Plains Indians during the same period, and Walker (a former commissioner of Indian affairs) was thoroughly aware of its centrality to any program of pacification (he is quoted to this effect in R. Athearn, *Union Pacific country* (Lincoln, NE, 1976), 206).

[25] F. Walker, "Instructions to assistant marshals," May 1, 1870, reprinted in Wright and Hunt, *History and growth*, 156.   [26] *Ibid.*, 168.

instructive to note [for the benefit of recalcitrant citizens] that at the Census of 1870 the agents of the Census in only two or three instances throughout the whole United States found it necessary to resort to the courts for the enforcement of the obligation to give information as required by the Census Act.[27]

The attention Walker devotes to this matter suggests that, even if only a handful of cases ever made it all the way to court, an initial refusal to give information must have been a relatively common occurrence. Another form of resistance was the deliberate giving of erroneous information. Walker's 1880 instructions address this at some length:

> It is further noted that the enumerator is not required to accept answers which he knows, or has reason to believe, are false ... Should any person persist in making statements which are obviously erroneous, the enumerator should enter upon the schedule the facts as nearly as he can ascertain them by his own observation or by inquiry of credible persons. The foregoing remark is of special importance with reference to the statements of the heads of families respecting afflicted members of their households ... It not infrequently happens that fathers and mothers, especially the latter, are disposed to conceal, or even to deny, the existence of such infirmities on the part of the children. In such cases, if the fact is personally known to the enumerator, or shall be ascertained by inquiry from neighbors, it should be entered on the schedules equally as if obtained from the head of the family.[28]

By requiring that enumerators be residents of their districts, and by reducing the size of districts to areas containing no more than 4,000 inhabitants, Walker could take a more effective advantage in 1880 of the local knowledge possessed by enumerators. Where that failed, he urged them to rely on the time-honored control strategy of asking neighbors to supply information.

His strenuously adamant response to a suggestion that census taking was an unconscionable invasion suggests that Walker must have been sensitive to the various ways in which Americans evaded or resisted enumerations:

> I do believe that the American people have outgrown the little, paltry, bigoted construction of the Constitution which, in 1850, questioned in Congress the right of the people of the United States to learn whatever they might please to know regarding their own numbers, condition and resources. It has become simply absurd to hold any longer that a government which has a right to tax any and all products of agriculture and manufactures, to supervise the making and selling of butterine, to regulate the agencies of transportation, to grant public monies to schools and colleges, to conduct agricultural experiments and distribute seeds and plant cuttings all over the United States, to institute scientific surveys by land and deep soundings at sea, has not full authority to pursue any branch of statistical information which may conduce to wise legislation, intelligent administration, equitable taxation, or in any other way promote the general welfare.[29]

---

[27] *Ibid.*, 169.　　[28] *Ibid.*

[29] F. Walker, "The Eleventh Census of the United States," *Quarterly Journal of Economics* 2 (1888), 135–161, reprinted in *DES*, vol. II, 79–80.

Walker here conflates "anything the people might be pleased to know about themselves" with "statistical information which may conduce to wise legislation, intelligent administration, equitable taxation, or in any other way promote the general welfare." It was in fact he (and James A. Garfield), heeding the advice of a relatively small number of reform-minded professionals, who enshrined the assumed congruence between the two kinds of information in the Census Act of 1880; "the people" had little direct say in the matter. His extreme annoyance can be read as a symptom of his awareness that "the people" were not actually the authors of this program of observation, and often reacted with distrust when they found enumerators at the door.

Certainly, groups having an interest in preventing an accurate count would make trouble, and in fact both Philadelphia and New York City were recounted in 1870 as a result of widespread objections to the first results. Much as the Lakota on the Great Plains resisted censuses during this period because they knew their rations of food and supplies from the US government would be reduced on a precise count, the major metropolises of the eastern seaboard understood that an accurate count would probably diminish their congressional representation.[30] More generally, "in the absence of definite information estimates as to the growth of cities and states soon become wild and extravagant. Cities vie with cities, and states with states, in their boasts of population and wealth, like individuals bidding against one another at an auction..." In particular (according to Walker), "the [New York City] authorities had ... committed themselves to the deepest hostility against the Census; and both the original enumeration and the re-enumeration under executive order were followed with eager and vindictive criticism, while every obstacle, short of actual physical resistance, was thrown in the way of the agents of the General Government."[31] Walker felt vindicated by the fact that in both cases, the re-counts yielded totals very close to the first enumeration.

It should be clear that access to information by census takers is a political matter, and the politics has been (and still is, in many places) closely related to the structure of "rule from a distance." Quite apart from the various time and place-specific political reasons people have had for withholding information, giving information to a stranger who will send it to a place where the individual can no longer control what is done with it, or what role it plays in decisions that rebound upon the giver, remains a potentially unsettling experience. It is not at all surprising that more or less organized resistance to censuses has been a recurring feature of "home" as well as of colonial rule.[32]

---

[30] M. Hannah, "Space and social control in the administration of the Oglala Lakota ("Sioux"), 1871–1879," *Journal of Historical Geography* 19, 4 (1993), 411–426.

[31] F. Walker, "Report of the Superintendent of the Census to the Secretary of the Interior," Dec. 26, 1871, in Office of the Census, *Ninth Census*, vol. I, xx, xxi.

[32] See Cohn, "The census, social structure and objectification," 238–239; Hannah, "Space and social control," esp. 418–426.

The spatial politics of the observational field: assortment

Taken together, reference grids, physical access and access to information construct an *abstract* field of observation, one with clear boundaries and subdivisions, across all of which the agents of the governmental gaze can travel without significant impediment, and throughout which they can expect to be provided with complete and accurate information. An abstract field forms the background and condition of possibility of *assortment*, whereby the *units* being observed (whether resources, people or activities) are unambiguously identified and distinguished from one another, and fixed to clearly defined locations. Not only census taking, but also "[t]racking property ownership and inheritance, collecting taxes, maintaining court records, performing police work, conscripting soldiers, and controlling epidemics were all made immeasurably easier by the clarity of full names and, increasingly, fixed addresses."[33] Walker's struggles to improve the census as a technology of observation involved not only the pursuit of *access* but also the battle to *assort* the population and its doings. That it was a battle can be seen in the following passage from an article written in 1891, where Walker compares the difficulty of census taking in New York City and Philadelphia:

In comparison with such a task [as enumerating New York City], a census of Philadelphia is child's play. There we have a city openly built, with ninety houses to every hundred families. Tenement houses are rare. Few of the people sleep in stables, in cellars, or in lofts. The houses are set squarely on the street. Four fifths of the inhabitants are native-born, and all, but a trifling percentage, of English speech. Merely to state these facts is, to one who knows anything of New York, enough to show the difficulties of enumeration in [the latter] city, which in 1880 had for 243,157 families but 73,684 dwellings, of which perhaps 20,000 were tenement houses, within the meaning of the Sanitary Acts. New York is a city with crowded and crooked courts, and alleys in the lower parts, and with thousands of shanties, sheds and inhabited sties in the upper parts . . .[34]

The 1880 enumerators did not venture into this epistemological morass utterly without preparation. Walker's general instructions to them included a passage, without precedent in the 1870 instructions, transparently aimed at the enumerators of New York City and other large urban cores:

By individuals living out of families is meant all persons occupying lofts in public buildings, above stores, warehouses, factories, and stables, having no other usual place of abode; persons living solitary in cabins, huts or tents; persons sleeping on river boats, canal boats, barges, etc., having no other usual place of abode; and persons in police stations having no homes. Of the classes just mentioned, the most important

---

[33] Scott, *Seeing like a state*, 71; see also D. Littlefield and L. Underhill, "Renaming the American Indian: 1890–1913," *American Studies* 12, 2 (1971), 33–45.

[34] F. Walker, "The great count of 1890," *The Forum* 11 (1891), 406–418, reprinted in *DES*, vol. II, 117–118.

numerically, is the first, viz: those persons, chiefly in cities, who occupy rooms in public buildings, or above stores, warehouses, factories, and stables. In order to reach such persons, the enumerator will need not only to keep his eyes open to all indications of such casual residence in his enumeration district, but to make inquiry both of the parties occupying the business portion of such buildings and also of the police.[35]

As the image of Philadelphia suggests, the most legible population is one perfectly assorted into nuclear families, each occupying a single, detached home at a fixed address. Walker subscribed to the orderly planning aesthetic shared by Le Corbusier, Julius Nyerere and countless other modern orchestrators of the urban and rural "rationalization" schemes diagnosed so thoroughly by James Scott in *Seeing like a state*. Scott identifies a few general features of this aesthetic: a preference for regular, geometric patterns best appreciated from a distant, bird's-eye perspective rather than from that of "street level"; a preference for clearly demarcated functional specialization (for example, business vs. residential areas) also most intelligible "from above" rather than on the ground; and a preference for standardized forms capable of being imposed on any concrete situation.[36] Like countless modern planners since, Walker tended to assume that those spatial arrangements most conducive to governmental observation "from above" were also those most conducive to social, moral and economic health.

Conditions like those described in Jacob Riis's famous *How the other half lives*, first published in 1890, were as far as possible from Walker's (and many other reformers') ideal of a perfect assortment, a one-to-one mapping between families and residences. Here is Riis describing what he found in "Jewtown" on the lower East Side:

It is said that nowhere in the world are so many people crowded together on a square mile as here . . . Here is [a tenement] seven stories high. The sanitary policeman whose beat this is will tell you that it contains thirty-six families, but the term has a widely different meaning here [than] on the avenues. In this house, where a case of small-pox was reported, there were fifty-eight babies and thirty-eight children that were over five years of age. In Essex Street two small rooms in a six-story tenement were made to hold a "family" of father and mother, twelve children, and six boarders. The boarder plays as important a part in the domestic economy of Jewtown as the lodger in the [Italian and Irish] Mulberry Street Bend. These are samples of the packing of the population that has run up the record here to the rate of three hundred and thirty thousand per square mile.[37]

[35] Walker, "Instructions to enumerators," May 1, 1880, 168.

[36] Scott, *Seeing like a state*, 107–111, 224–225, 237–238.

[37] J. Riis, *How the other half lives* (New York, 1957 [1890]), 77–78. For Riis, the "other half" is composed chiefly of immigrant ethnic groups. Like Walker and many of their contemporaries, Riis strongly implicated immigrants in the whole range of urban pathologies, though "native" white Americans undoubtedly formed a significant proportion of poor tenement dwellers. In Chapter 6, we will see Walker become increasingly obsessed with immigrants as the root of many social ills.

Riis's account portrays tenement populations as ill-assorted not only because they departed so markedly from the ideal of one nuclear family to a residence, but also because they moved around in ways unknown and invisible to outsiders. His description of the network of sewers and cellars must have been particularly alarming to Walker and other representatives of the governmental gaze:

The fact was that big vaulted sewers had long been a runway for thieves – the Swamp Angels – who through them easily escaped when chased by the police, as well as a storehouse for their plunder . . . The flood-gates connecting with the Cherry Street main are closed now, except when the water is drained off. Then there were no grates, and it is on record that the sewers were chosen as a shortcut habitually by residents of the court whose business lay on the line of them, near a manhole, perhaps, on Cherry Street, or at the river mouth of the big pipe when it was clear at low tide.[38]

Such sanctuaries added a whole subterranean dimension to invisibility, and would have compounded the difficulty of census-taking immeasurably (in both senses of the word). The overall effect of such glaring lack of assortment on the prospects for effective governmental observation is brought out symbolically in Riis's description of "the Bend": "The whole district is a maze of narrow, often unsuspected passageways – necessarily, for there is scarce a lot that has not two, three, or four tenements upon it, swarming with unwholesome crowds. What a birds-eye view of 'the Bend' would be like is a matter of bewildering conjecture."[39]

The difficulty of pinning down units of enumeration in such a situation was partially dealt with by definitional fiat in Walker's instructions to enumerators for 1870, where "by 'family' (Column 2) is meant one or more persons living together and provided for in common. A single person, living alone in a distinct part of a house, may constitute a family; while, on the other hand, all the inmates of a boarding house or a hotel will constitute but a single family, though there may be among them many husbands with wives and children." The 1880 instructions had the potential to confuse what for the enumerators must already have been a complicated judgment by declaring that persons "living out of family" can still be "families" if they sit down to eat at "separate tables."[40] The effect of these instructions on aggregate figures of family size in major cities (indeed, for the nation as a whole) is interesting to contemplate, especially in light of Walker's preoccupation with fertility rates as indicators of social health (see chapter 6). By themselves, such instructions could do little to improve the assortment of the population.

A second major obstacle to assortment, which had the effect of compounding the problem of mobility, was somewhat more tractable. This was what Walker called in his 1871 Report to the secretary of the interior "the essential

---

[38] *Ibid.*, 27–28.   [39] *Ibid.*, 43.
[40] Walker, "Instructions to assistant marshals," 156–157; "Instructions to enumerators," 171.

viciousness of a protracted enumeration." However well enumerators could overcome the barriers in the way of distinguishing units of enumeration, the degree of fixity they could assume for these units rapidly diminished as the count extended over weeks and months.

When it is considered how many thousands of persons in every large city, how many tens of thousands in a city like New York, not only live in boarding houses, but change their boarding houses at every freak of fancy or disgust, not to speak of those who leave under the stress of impecuniosity and therefore are not likely to leave their future addresses or advertise their residence, it will be seen how utterly unfitted is such a system of enumeration to the social conditions of the country at the present time.[41]

The 1850 law under which the 1870 Census was taken originally allowed assistant marshals five months, from June 1 to November 1, to complete and return their schedules, with extensions possible for up to two additional months. Strenuous lobbying from Walker and Garfield persuaded Congress, though it would not pass a new census act, to amend the time limit in the 1850 law for the 1870 count. This reduced the window to three months and nine days, with possible extension in special cases up to a total of four months (from June 1 to October 1). This modest improvement did not satisfy Walker, as it left too large a window not only for tenement dwellers but for the annual summer peregrinations of the growing middle and upper classes.[42] The 1880 law shortened the window to one month, and Walker tended reluctantly to agree with critics who claimed that the ideal of a census taken in a single day was probably unreachable for the United States.[43] In the 1872 Report, he admitted that "there are undoubtedly regions in which [a one-day] enumeration would require that nearly every man should be commissioned as an assistant marshal for his own family."[44] A comprehensive mail-in census would, in effect, commission the whole population, but, at that time, Walker was only just beginning to see the potential of such methods, and mentioned the idea only in passing.[45] The reduction from a maximum enumeration district size of 20,000 to one of 4,000 between 1870 and 1880 certainly allowed a quicker count, but in parts of the west, 4,000 people might be scattered across hundreds or even thousands of square miles.

Having reviewed the ways in which Walker understood and tried to improve the abstraction and assortment of the field of observation, it is worth pausing to ask about the relationship between these two conditions. In one important sense, it is a relationship of contradiction. Although abstraction is a condition of possibility for assortment, it also constitutes a limitation on it. Specifically,

---

[41] Walker, "Report of the Superintendent of the Census," 12–26–1871, xxii.
[42] F. Walker, "The United States Census," *The Forum* 11 (1891), 258–267, reprinted in *DES*, vol. II, 106.   [43] Wright and Hunt, *History and growth*, 941.
[44] Walker, "Report of the Superintendent of the Census," xxii.
[45] Walker, "Defects of the census," 57.

the ever improving transportation infrastructure that played such an important enabling role in opening up the field of governmental observation to
census takers, police and other officials at all levels of government, also made
it easier for the population to move around in search of work or opportunities
of other sorts. This is one important respect in which governmental power
differs from panoptic power. Panoptic power is premised on a technology of
vision that allows authorities to see without being seen. By contrast, an act of
observation associated with a census requires that the agents of vision travel
to their objects using the same infrastructures available to the objects themselves. The easier it is for government agents to move about, the easier it is likewise for the population at large.

The wholesale transformation of American mobility during the late nineteenth century, which led to the situations described by Jacob Riis, hardly
needs more than the briefest rehearsal here.[46] As industrialization, immigration and urbanization went into high gear, the proportion of the American
population on the move *as* individual job seekers skyrocketed. Their destinations were the major urban hubs, especially New York City. As Riis put it in
1890:

The metropolis is to lots of people like a lighted candle to the moth. It attracts them
in swarms that come year after year with the vague idea that they can get along here if
anywhere . . . Nearly all are young men, unsettled in life, many – most of them, perhaps
– fresh from good homes, beyond a doubt with honest hopes of getting a start in the
city and making a way for themselves.[47]

Many struggling urban families (of all ethnicities) were willing to put up
boarders, and many landlords were able to make a killing by packing as many
working poor as they could into tenements. The arrival of millions of immigrants on the eastern seaboard caused this fluid population to explode, and
intensified and accelerated the growth of tenement districts and slums. As we
shall see in Chapter 6, Walker could not very well object to the increased
mobility of the population, however difficult it made census taking. He held
the mobility of workers to be at the root of a healthy American political
economy.

The contradiction between abstraction and assortment was in effect a limit
on the possibility of complete knowledge. "If, by reason of improved facilities
for travel, and the greater restlessness of our population, an increasing number
shall escape enumeration at each successive census, it is a matter over which
the Census Office, as presently constituted, has little control."[48] Yet, clearly,

---

[46] An overview of the processes involved can be had from D. Ward, "Population growth, migration, and urbanization, 1860–1920," in Mitchell and Groves, *North America*, 299–320. For a
more extended treatment, see D. Ward, *Cities and immigrants: A geography of change in nineteenth century America* (New York, 1971).   [47] Riis, *How the other half lives*, 60.
[48] Walker, "Report of the Superintendent of the Census," Dec. 26, 1871, xxxii.

Walker's attitude was never passive. Like the forms of resistance mentioned earlier, Walker proposed to overcome the limits posed by mobility through recourse to local knowledge. Whether the problem was mothers ashamed of their childrens' condition or floating populations of drifters living in lofts, Walker urged his enumerators to rely on their own familiarity with their districts, or to ask business owners, police, neighboring families, in short anyone who seemed trustworthy and had some information. Again, Walker was very cautious, deliberate and "governmental" when it came to the question of regulating the social body (see Chapter 7), but epistemologically speaking, at the moment of observation, he was as "imperialistic" as any colonial administrator.

The other way in which Walker tried to get control of the contradiction between fixity and mobility was to map it onto a gender division of labor in which men were properly mobile (the agents of abstraction) and women properly fixed (the anchors of assortment). Ideally, as long as women remained at home with distinct (enumerable) nuclear families, men could enjoy the mobility necessary to help the government count the population, and to maintain a healthy political economy (see Chapter 6). The masculinity of the agents of abstraction is fully in line with the gendering of social sciences noted by feminist scholars: "The visual has always been central to masculinist claims to know . . . seeing was certainly important to the emergence of the social sciences toward the end of the nineteenth century."[49] What Gillian Rose writes of present-day time geographers applies equally well to the enumerators and supervisors of the 1880 Census: "Their masculine consciousness peers into the world, denying its own positionality, mapping its spaces in the same manner in which white Western male bodies explored, recorded, surveyed, and appropriated spaces from the sixteenth century onwards: from a disembodied location free from sexual attack or racist violence. Space for them is everywhere; nowhere is too threatening or too different for them to go." By contrast, women are "natural objects," passive and immobile in their domestic settings.[50] It was no accident that Walker claimed impartiality for acts of observation through the medium of a gendered rhetoric:

Because women invade the forum, and crowd us from our places on the public platform, shall we, therefore, take refuge in the kitchen, or be so base as [to] seem to know what passes in that realm of blackness and smoke? Perish the thought! The object of this paper is to present facts that are not of personal experience . . . facts gathered by thousands of men, who had as little notion what should be the aggregate import of their contribution, as my postman has of the tale of joy, of sorrow, or of debt, which lies snugly folded in the brown paper envelope he is leaving this moment at my door. No momentary fretfulness of a mistress overburdened with cares; no freak of insolence

---

[49] G. Rose, *Feminism and geography: The limits of geographical knowledge* (Minneapolis, MN, 1993), 38–39.   [50] *Ibid.*, 39; see also Rose's chapter 5.

from a maid elated by a sudden access of lovers; no outbreak of marital indignation at underdone bread or overdone steak, can disturb the serenity of this impersonal and unconscious testimony of the census.[51]

All trace of the instructions to obtain information through personal or local knowledge (which would often have meant reliance on the knowledge of neighborhood women) has been expunged from this idealized image of the intrepid male enumerator. This gendered spatial division of labor, as I suggested above, informs Walker's understanding of how American society should be, not merely how it would be most legible. Again, as James Scott points out, governmental authorities have long tended to conflate legibility and social health. I will pick up this theme again in Chapter 6.

The spatial politics of centers of calculation: Centralized control

Implicit in many of Walker's actions as superintendent was a spatial politics of centralized control. He strove not only to organize the field of observation in terms of abstraction and assortment, but also to gain and retain as much control as possible over the enumerations from his office in Washington. A national system of governmental knowledge, like "rule from a distance" more generally, is most effective if anchored to a central authority able to orchestrate the mastery of territory. The "rays of vision" which play over the surface of the field of observation must converge somehow at a central "eye." Matthew Edney, in his study of the British mapping of India, provides an extremely useful distinction around which to organize a discussion of this issue: the distinction between central control of *acts of observation,* central control of *observers* and central control of *information.*[52] Together these dimensions of central control complete the basic architecture of governmental vision. None of these three types of central control could be taken for granted, either by the East India Company in the early nineteenth century, or by the superintendent of the US Census fifty years later. Here again, as with the organization of the field of observation, Walker was able to register some advances in the 1880 Census relative to the 1870 count.

In a sense, central control of acts of observation simultaneously makes information more amenable to central control. This dual goal can be seen in Walker's efforts at the standardization of the procedures whereby assistant marshals (and then enumerators) filled out schedules. In 1870, the instructions to enumerators were far more elaborate than they had been under Walker's predecessor Joseph C. G. Kennedy in 1860, despite the fact that the two counts were taken under substantially the same law. Walker's instructions comprised a densely typed ten-page pamphlet, which included not only guidance on how

---

[51] F. Walker, "Our domestic service," *Scribner's Monthly* 11 (1875), 273–278, reprinted in *DES,* vol. II, 225.   [52] Edney, *Mapping an empire,* 138–139, 163.

best to approach people, but far more exhaustive rules for classifying people and their occupations, and provisions for regular reporting of activities during the enumeration. The instructions open with an insistence that assistant marshals "always carry the full pamphlet of instructions," and go on to devote pages of minute guidelines, especially for determining the categories of "occupations." Alarmed, on perusing the 1860 results, to find, for example, that while Illinois had 554 bookkeepers, Massachusetts, 593 and Pennsylvania, 519, New York had *none*, Walker hammered away in the 1870 instructions at the theme of maximum possible specificity: "Describe no man as a 'mechanic' if it is possible to describe him more accurately . . . Distinguish between stone masons and brick masons," etc.[53] In 1870, for the first time, assistant marshals were required to report to marshals on their activities once every fortnight; in 1880, Walker had special postcards printed which enumerators were to use for *daily* reports. Enumerators were to submit two postcards a day, the gray colored card to the local supervisor and the buff colored card to the Census Office in Washington. Each card had printed on the back a basic form with blank spaces for number of "persons, farms, etc." counted that day, as well as a statement of the time "actually and necessarily occupied in this service."[54] After completing the enumeration, the enumerator was further required in 1880 to advertise and hold a two-day session at the local courthouse for the purpose of correcting any errors. This was to be run like a hearing, with witnesses giving formal testimony.[55] The correction hearing served two purposes. The first of these was to minimize the need for subsequent investigations by the central office. In 1870, Walker had included in his instructions to assistant marshals the following warning: "When the Census Office is put to trouble and expense, by having to obtain through subsequent correspondence the answers to [mandatory] questions, the cost of clerk hire and correspondence to the Department will be estimated, and deduction will be made [from assistant marshal pay] for work not done."[56] The second, related purpose served by the hearings was to place as much of the burden for objectivity as possible in the field, with the observers and their acts of observation, rather than with the central office (see below).

Though generally satisfied that such measures improved the observational competency of his proxies in the field, and despite his professed interest in a timely count, Walker was anxious about the apparent haste of his enumerators. The Census Office sent approximately 900 letters to enumerators during the second half of 1880, almost all concerning details of their compensation. Of these, probably at least one third included the following paragraph, which Walker apparently formulated for his correspondence clerks:

[53] F. Walker, "American industry in the Census," *Atlantic Monthly* 24 (1869), 689–701, reprinted in *DES*, vol. II, 21; Walker, "Instructions to assistant marshals," May 1, 1870, 159.
[54] Walker, "Instructions to assistant marshals," 155; Walker, "Instructions to enumerators" 175.
[55] *Ibid.*, 176–177.   [56] Walker, "Instructions to assistant marshals," 155.

As the number of names taken in that time is —, the number of farms — and the number of names on Schedule 5, —, it occurs to me to suggest, in your own [pecuniary] interest, whether there may not be some error in regard to your report of time, as it seems to be quite short for the work done. Should any such error exist, this office will cheerfully allow you to make the necessary correction.[57]

Walker remained uneasy about the quality of the acts of observation being carried out all over the country, in part because of the relationship between the size of enumeration districts, the qualifications of enumerators and the character of the inquiries they made. Even with the 1880 law, census taking involved a tradeoff in which the *areal extent* of each district was minimized, but at the price of enumerators being responsible for collecting a *wide variety of types of information*. Walker realized that, however necessary this arrangement might be, given the limited willingness of Congress to devote funds to the census, it constituted a threat to standardization. The comparability of statistics on, say, mortality, *over a wide geographical area*, would require that a large number of enumerators, each responsible for precise collection of many different kinds of data, would nevertheless record mortality in exactly the same way. What Walker had in mind as an alternative can be glimpsed in his response to inquiries from Congress in 1875 about whether it would be feasible or worthwhile to take a special census in that year so that the results could be incorporated into the centennial celebrations of 1876. There, along with a streamlined count, he recommends that Congress agree to fund the hiring of experts to serve not merely as consultants on interpretation and compilation (a practice adhered to by superintendents since mid-century), but as the actual enumerators of economic statistics. His plan was to assign each expert a much narrower set of topics, but a much larger district, so that, for example, the metalworking industries of an entire city would have been enumerated by a single expert, making for more informed observational acts which would yield information whose comparability could be assumed to extend over a wider area.[58] Walker had only limited means with which to engage experts, especially in 1870, but he did appoint various deputy marshals to oversee economic and social statistics.

From surviving correspondence with his friend Daniel Coit Gilman, it is possible to get a sense of how Walker understood expertise. Gilman would go on to become the first president of Johns Hopkins University and have a distinguished career as an educational reformer. Before that, Walker enlisted him as the special deputy marshal to oversee the collection of the social statistics of Connecticutt. Social statistics (pauperism, crime, education, libraries and

---

[57] National Archives and Records Administration [NARA], Record Group 29, Records of the Bureau of the Census, Administrative Records of the Census Office, Entry 69: "Copies of letters to enumerators, 1880."

[58] F. Walker, "Defects of the Census of 1870," from "Report of the Superintendent of the Census," Nov. 1, 1874, House Executive Document 1, part 5, vol. 6 (43rd C., 2nd S.), reprinted in *DES*, vol. II, 55–56.

other institutional phenomena) were enumerated on Schedule 5 during 1870, the only population schedule not requiring assistant marshals to visit every household. What stands out in Walker's instructions to Gilman is the wide leeway given to the latter's judgment, and the implicit trust Walker placed in "sagacious men." For example, Walker gently chides Gilman for not including estimates where actual figures were unavailable. Or, "in the matter of the distribution of the persons in prison on the first of June, 1870, with respect to nativity and color, I will be satisfied with a generalization from any single case which you believe would naturally afford a fair average experience in these respects."[59] For Walker, expertise resided not so much in specialization as in the virtues of an upright, well-educated *man*. This way of thinking also informed his methods for organizing the central office, as we shall see below.

Interestingly, one of the ways in which Walker here recommends streamlining a hypothetical 1875 count is by having experts look only into "the gigantic establishments in which hundreds of workmen are employed with books kept by double entry," and leaving aside small shops "in which single artisans work at their trades, and perhaps chalk their accounts on the wall." In other words, enumeration of manufacturing should be "restricted to industries which are carried on in large establishments, and which, consequently, it is possible to enumerate with completeness and accuracy."[60] Acts of observation could be improved not merely by engaging experts to do the looking, but by concentrating on the most *legible* units.

Control of acts of observation was clearly entangled with the issue of control over who did the observing, and it was on this issue that Walker fought his most explicitly political battles. He never squandered an opportunity to explain to Congress or the general public how debilitating it was to be forced to rely on the entrenched patronage system as the pool of labor for field observers. The 1850 law under which the 1870 Census was taken provided that US marshals oversee the enumeration locally, and do the hiring of the assistant marshals who would actually collect the information. This put the observers under the control of the party-run patronage systems rather than the Census Office, and indeed it was to protect this patronage system that US marshals had demonstrated in Washington DC, and that the Senate had (partly as a result) refused to pass a new Census bill for 1870.[61] Walker repeatedly stressed that not only did the marshals have lots of other duties to attend to besides supervising a census, they (and the assistants they appointed as enumerators) for the most part had no special experience with or interest in census taking, even if they were honest and sincere in their efforts. Perhaps most galling to Walker was his complete lack of control over either hiring or firing. This situation almost ensured highly uneven information quality.

[59] Walker to Gilman, May 10, 1871, Daniel Coit Gilman Papers, Special Collections, Johns Hopkins University, MS 001. [60] Walker, "Defects of the Census of 1870," 52, 53.
[61] Walker, "Enumeration of the population, 1870–1880," 62–63.

The law for the taking of the 1880 Census finally established central control over the observers. Instead of 60 US marshals, 150 supervisors, drawn from all regions of the country, hired by Walker and installed in their positions full time for the duration of the count, were charged with recommending enumerators on the basis of competency and interest. The enumerators themselves, hired without regard for party affiliation (and subject to review and dismissal by the central office), were often school teachers, county clerks, tax assessors "or other persons having familiarity with figures and facility in writing."[62] In urging funds to hire as many supervisors as possible, Walker was simply reapplying a strategy of control that had proven successful for him during his brief stint as commissioner of Indian Affairs in the early 1870s. As described by Carroll D. Wright in his 1897 Memorial Address for Walker at MIT, this strategy is clearly of more general use in programs of national control. Upon taking the helm at the BIA, Walker

> soon discovered that while this was a branch of Government having its business widely scattered, and with most delicate and responsible duties assigned to untried men at remote agencies in the West, there was no system by which the affairs of these agencies were inspected and properly supervised by officers directly responsible to the central office at Washington . . . This need General Walker temporarily supplied by the appointment of 'special commissioners,' who visited as inspectors the places of most importance and interest, and reported directly to him. Congress agreed with his recommendation that such supervision be institutionalized, and the office of United States Indian inspector was promptly created.[63]

The elaboration of observational procedure and control over personnel would have been easier and more efficient had the central office been a permanent bureaucracy rather than an operation conjured up *de novo* every ten years. A permanent bureau would have engaged a more permanent professional staff independent of and much better insulated from party politics; it would have constituted a repository for institutional memory in the form of records from which each successive enumeration could learn; it would have allowed a range of different enumerations to be undertaken on a more continuous basis; and it would have enabled much more thorough training and preparation for each of the major enumerations. In short, it would have constituted a stable "eye" for the governmental gaze, a "center of calculation" to anchor the governmental "network." These arguments were made repeatedly by Walker and others connected in one way or another with social statistics, but it was not until 1902 that a permanent Bureau of the Census was finally established.[64]

---

[62] *Ibid.*, 64.
[63] C. Wright, "Francis Amasa Walker," *Publications of the American Statistical Association*, new series 38 (1897), 267.
[64] See, for example, the section entitled "Defects of the Census of 1870," in Walker, "Report of the Superintendent of the Census" Dec. 26, 1871.

The spatial politics of centers of calculation: Objective compilation

In the absence of a permanent institutional house, Walker nevertheless did all he could to rationalize the temporary one. The role of the central office in producing a plausibly "objective" census was not confined to controlling acts of observation and observers in the field. If anything, its main task was to *compile* information in a timely, efficient and accurate manner, and this too involved a subtle sort of politics. Here again, Matthew Edney's study of the British mapping of India is immensely helpful in sorting out the issues. Edney raises the crucial question of precisely which part of an undertaking as elaborate as the mapping of India or the taking of a census is the seat of its purported "objectivity." He traces a fascinating change in the thinking of the British, from the assumption that the burden of objectivity rested with the *compilers in the central office* to the idea that it rested with the *agents of observation in the field*. In the earlier understanding,

the data which contributed to the maps were indeed the result of measurement and observation, but they could only achieve greater significance and meaning within a graticule of meridians and parallels. In contrast, triangulation [which came increasingly to be seen as the only proper foundation for the mapping of India] provides its own framework whose authority derives entirely from its constitutive acts of measurement and observation. The basis of the map's cultural authority thus shifted [by the 1840s] from the cartographer in his office to the surveyor in the field.

Eventually the British "achieved a compromise which cloaked the continuing exercise of map compilation in the authority of systematic field observation."[65]

The situation with the United States Census was somewhat different, in that the role of compilation in the central office was not associated with such a well-defined activity as "cartography," and thus was probably not at any time considered by the general public to be such a well-defined and distinct "step" in the process of completing a census. Compilation was probably considered by most non-statisticians to be a matter merely of "checking the figures and adding them up." Like map compilation, it was not supposed to add anything to or subtract anything from the information coming in, but simply to aggregate it and reduce it to a manageable level of complexity for publication. The "cultural authority" of the census, its objectivity, originated in the acts of observation in the field. More than cartography even, census taking exemplified just the sort of empiricist approach to observing the world, centered on individual facts, that Walker championed so strenuously in economics and more generally. Thus, the more quickly and transparently the Census Office could process the facts, the easier it would be to preserve the perception of objectivity. If compilation could be made sufficiently unobtrusive, it would be

---

[65] Edney, *Mapping an empire*, 30.

easier to assume that "the American nation" was a reality, not a construct.[66] This unobtrusiveness is the counterpart, at the scale of national governmentality, of the ideally "automatic" and impartial nature of observation in the panopticon. In both cases, an important goal of "power/knowledge" is to convince the objects of observation, whether individual prisoners or a national population, that there is nothing political going on at the central point of convergence of the information used to govern them.

To make compilation as fast and efficient as possible, Walker introduced two new kinds of order to the central office: a competency examination for prospective clerks and a comprehensive system of record-keeping oriented to the efficient use of time by the clerks he hired. These were but two components of a new culture of record-keeping and accountability with which Walker infused the Census Office.[67] As far as the geographic distribution of the personnel who would be subject to this new culture, Walker's hiring policies were as impartial as conditions would allow. Because the two national political parties remained the only organizations through which Walker could have access to a national labor pool, he had no choice but to rely on them. The recommendations he received for applicants thus read very much like patronage recommendations, with some applicants able to marshal the imprimaturs of three or more representatives or senators, and a significant number bearing pleas from as eminent a personage as Rutherford B. Hayes.[68] However, Walker hired impartially from both parties, and the geographical pattern of his hirings seems roughly to have followed a distance-decay function centered on Washington DC and adjusted for population density. For the 1870 Census, by my calculation, 29 percent of the 576 administrative employ-

---

[66] See Anderson, *Imagined communities*, pp. 164–170.

[67] The administrative records surviving in the National Archives are far from complete (one of the effects of the lack of a permanent Census bureau), and it is thus difficult to be precise about how much more record-keeping there was under Walker's reign than under that of his predecessor Joseph Kennedy. Nevertheless, it is safe to say that there *was* more under Walker. Not only are the surviving records from the Ninth and Tenth Censuses far more numerous than those from the Eighth, they are also far more *specialized* in their topics. Where one "List of Census Office employees, 1860–1863" (NARA Record Group 29, Records of the Bureau of the Census, Administrative records of the Census Office, Entry 31) encompasses the names, states of residence, nature of work and supervising clerks for all central employees, five different personnel books survive from the 1870 count, each dealing with a different kind of personnel record (alphabetical list of employees; journal of personnel actions; "personnel ledger"; index to journal and ledger; and "time and service record of census office clerks"). The number of personnel-related books surviving from the 1880 count is ten, even more specialized than in 1870 (NARA *Preliminary inventories No. 161: Records of the Bureau of the Census*, K. Davidson and C. Ashby (compilers), (Washington, DC, 1964), 22–27). What distinguishes the 1870 and 1880 records from the one book surviving from the Eighth Census is that whereas Kennedy merely used ledger books with blank, unruled pages, Walker had the leaves of each different book printed as customized forms, with elaborate rulings and headings specific to a particular way of organizing personnel information.

[68] NARA Record Group 29, Entry 45: "Record of recommendations for applicants, 1880–1881," 3 vols.

ees listed for the period 1870 to 1872 were residents of Washington, DC; 26.5 percent residents of the Middle Atlantic states to the north (Maryland, Delaware, Pennsylvania, New York and New Jersey); 18.4 percent from the (trans-Appalachian) Midwest; 12 percent from the Southeast (the lion's share of these from nearby Virginia); just under 10 percent from New England; and much smaller contingents from the West, the Plains and the Southwest.[69] This pattern was typical for most Washington bureaucracies at the time.[70] Particularly the modest contribution of New England is a sign that Walker was not out to "reward his own."

By contrast, the geographical pattern shown by Kennedy's central office employees for 1860 to 1863 departed significantly from a distance-decay distribution modified for population density: Washington, DC and the Middle Atlantic states each contributed roughly 22 percent of employees; the Southeast contributed 9 percent; and New England, 8 percent. The anomalies were, first, that Pennsylvania accounted for an inordinately large proportion of the Middle Atlantic contingent, and second, that the·Midwest sent by far the largest group, at 33 percent of the total.[71] These curiosities cease to be so curious with the knowledge that Kennedy was a native of Meadville, PA, a town on the western side of the Appalachians near Lake Erie and the Ohio border (and therefore well connected with all midwestern states bordering the Great Lakes). His original appointment to head the 1850 Census (which eventually led to his being reappointed in 1860) had come as a reward from Whig President Zachary Taylor, in recognition of Kennedy's efforts to get out the Whig vote in Meadville.[72] At least in hiring policy, Kennedy continued the patronage tradition from which he himself had benefited.

In the interest of guaranteeing a minimal level of competency, Walker subjected applicants for clerk positions to a standardized examination covering basic writing and arithmetical skills. He was so pleased with the results of this exam in 1870 that he devoted most of the section on "Census Office organization" in his 1872 report to the secretary of the interior to an account of it. Although I have not been able to locate a copy of the exam itself, it apparently was not easy. Scores could range from 0 to 1,000, and more than half of the 719 applicants who took it scored below 500. The peak of the distribution of scores was between 400 and 500, and frequencies drop off rapidly in the higher intervals. Only six of the 719 applicants scored above 900, as compared with 52 between 0 and 100.[73] Only 401 applicants achieved a passing score, which Walker defined as 400 or better. The 1880 applicants took a similar exam.

[69] NARA Record Group 29, Entry 35: "List of Census Office administrative employees, 1870–1872," 2 vols.

[70] C. S. Aron, *Ladies and gentlemen of the civil service: Middle-class workers in Victorian America* (New York, 1987), 43.   [71] NARA Record Group 29, Entry 31.

[72] M. Anderson, *The American census: A social history* (New Haven, CT, 1988), 36.

[73] Walker, "Report of the Superintendent of the Census" Dec. 26, 1872, xlviii.

Not only competency, but *time*, was of the essence if the process of compilation was to appear routine and uncontroversial. Time was of course a key consideration from the perspective of Congressional funding for census work, but this was clearly not the only consideration driving Walker's emphasis on it. He gave the efficient use of time a prominent place in the personnel records. The "Personnel ledger" for 1870–72 gave each employee half a quarto page of space, dominated by a table of time categories. Most of these categories were particular kinds of "absent time," which was divided first into "with leave" and "without leave" and recorded by quarter. Absence with leave had to be identified as due to one or the other of five sub-categories: "election," "special," "vacation," "sick" or "without pay." Absence without leave could be classed under "sick," "excused" or "not excused." To the right of these columns were two others designated "net loss" and "net gain" (though it is not clear what could account for net gains in time other than working extra hours).[74] The "Time and service record of Census Office clerks" for 1870–71, which is mostly devoted to recording the substantive tasks employees were engaged in, also includes prominently placed columns concerning time lost, notably "time lost by unseasonable arrival or departure" and "time lost between arrival and departure."[75] The "Personal Record" for employees covering the period 1879 to 1882 is structured similarly to the "Personnel Ledger" of ten years earlier.[76] All of this was undoubtedly not merely a result of Walker's record-keeping mania, but also an attempt to keep employees constantly aware of the need for as much speed as the imperative of accuracy would allow.

Together with the examinations, the constant focus on time loss allowed Walker to claim with confidence, in his 1872 report to the secretary of the interior, that he had "fix[ed] a high standard of clerical efficiency, and exact[ed] the utmost of daily performance that could justly be required of the clerks of the office."[77] For the 1880 Census, of course, this was doubly true, but despite the rigid time discipline, the sheer size and complexity of the 1880 Census meant that compilation took up three years, and publication stretched over the better part of a decade (see Chapter 2). Despite the great campaign to appear impartial, the activities of the central office during compilation were not at all passive with respect to the incoming information. We can glimpse the active role of the central office in surviving correspondence between Walker and Daniel Coit Gilman from 1871. Though originally assigned responsibility for social statistics, Gilman ended up submitting statistics also on the value of real property owned by Connecticut citizens. In one letter, Walker explains that the experienced clerks in the central office would insist on a revaluation if they received figures which they considered too low based

[74] NARA Record Group 29, Entry 33: "Personnel ledger, 1870–1872."
[75] NARA Record Group 29, Entry 36: "Time and service record of Census Office clerks, 1870–1871."   [76] NARA Record Group 29, Entry 51.
[77] Walker, "Report of the Superintendent of the Census" Dec. 26, 1872, xxxvii.

on their own estimates of actual values. In a later letter, Walker thanks Gilman for writing a popular article on the pitfalls of taking the Census, and expresses his particular gratitude for Gilman's account of this pattern of undervaluation in property estimates. Walker writes, "the change made in respect to Vermont is just one of the kind which you indicate in the text of your article, the marshal having at first returned the state at an unreasonably low valuation, and having been persuaded, with some difficulty, to advance his figures to meet the views of the office."[78] While the source of the Census's objectivity was supposed to be *the enumeration itself*, the process of *compilation* played a major role in shaping the final figures.

Not only the management of information, but also the management of personnel in the central office showed interesting departures from the image of fully modern rationality, transparency and impartiality presented to the public. These departures related to gender. First, Walker hired a substantial number of women to work in the Census Office, at least potentially complicating the cultural association between maleness and "objective" scientific observation. The numbers for the 1870 count are not clear, since most employee lists give only first initials and no titles. However, the "Personal Record" of employees working on the 1880 Census does include the titles "Miss" or "Mrs." for almost all female employees, and thus provides a profile of gender and marital status. Out of a total of 1705 employees listed, 771, or 45.2 percent were women, a significantly higher percentage than were in federal government service as a whole at the time.[79] Since the Washington bureaucracies of the 1870s and 1880s were the national "beachheads" for women's entrance into white-collar work, these numbers mean that the Census Office was truly in the vanguard of the national trend. Women were hired, except perhaps for some of the locals living in the Washington, DC area, through the same party organizations that Walker used to hire men, and the geographical distribution of their states of residence is similar to that of the men. Of the 771 women, 520 were unmarried, somewhat above the overall proportion of roughly 2/3 unmarried for the Washington bureaucracies.[80] But marital status does not appear to have had much of an effect on geographical distribution of states of residence. It is difficult to learn much about how these women were understood and treated by Walker and his chief clerks, all except one of whom were men. But the clearly gendered conception Walker had of census work (and public life in general) makes it difficult to imagine that he was as impartial about gender as he was about party affiliation, that he was as comfortable

---

[78] Walker to Gilman, Mar. 20, 1872, Daniel Coit Gilman Papers.

[79] NARA Record Group 29, Entry 51; Aron, *Ladies and gentlemen of the civil service*, 5.

[80] Aron, *Ladies and gentlemen of the civil service*, 46, though Aron notes that cultural norms stigmatizing married women who worked may have prompted an unknown number to pose as unmarried *ibid.*, 50–51) Until the 1890s, unmarried women working in the Census Office were expected to resign in the event of marriage, though some did not (endnote 45, p. 208).

hiring women as he was hiring men. Cindy S. Aron shows that many women applying for work with government bureaucracies, well aware that they were seen to represent a challenge to the prevailing domestic ideology, employed a rhetoric of embarrassment and exaggerated passivity in explaining why they were forced to seek jobs outside the home.[81] It could well be that young women with writing and arithmetical skills simply made up a large part of the labor pool available to him.

The uniformly lower pay that women received in the federal bureaucracies may also have eased Walker's discomfort. The Census Office was always in dire budgetary straits, its impermanence as well as the potentially explosive political implications of its work on Congressional reapportionment leaving it more vulnerable than other branches to the whims of Congress. After 1870, it had become common practice for women to move upwards into "male" pay grades after beginning at the lower women's entry grades, but the savings to the Treasury were still substantial.[82]

One intriguing clue as to how Walker dealt with the presence of so many women appears in the surviving "Weekly service reports" covering early 1880, when the Census Office was just getting off the ground in preparation for the summer enumeration. Of fifty-six employees hired at that stage, twenty-one were women, and one of these, a Mrs. Kearfott, is listed as "superintending and directing work, ladies department."[83] The women were apparently segregated despite the fact that they were engaged in the same range of tasks as were the men. I have not been able to discover whether this gender division held through the whole period of central office activity, but in any case it appears to contradict Cindy Aron's claim that supervisors within federal offices at the time "either had to let women perform the same work as men, or reorganize and subdivide the work within their offices so that men and women carried out different jobs."[84] Since throughout the 1870s and 1880s the heads of executive departments still had complete freedom to run their offices as they saw fit, it is in any case clear that Walker did not see his employees through gender-blind lenses.

A number of fascinating issues are raised by this segregation of women, but I would like to concentrate on one in particular: the fact that Walker divided the central office into divisions on a basis other than the modern functional division of labor. There was a divisional structure in 1870 and in 1880; however, the divisions were designated not by the particular tasks assigned to them, but by the names of the chief clerks in charge. At least in 1870, and possibly also in 1880, Walker had each division working on different tasks at different times, and in many cases, different clerks within a single division would be differently engaged at the same time. It was not until the Eleventh

---

[81] *Ibid.*, 58–59, 101.   [82] *Ibid.*, 71.

[83] NARA Record Group 29, Entry 79: "Weekly service reports, January to April, 1880."

[84] Aron, *Ladies and gentlemen of the civil service*, 72.

(1890) Census that the divisions were clearly demarcated according to non-overlapping tasks. This was not unusual in the federal bureaucracies, but it is interesting, because in Walker's case, it could be read as a specifically *military* approach to organization (see Chapter 4). His divisions, like those in an army, were more or less interchangeable, and the focus was on the *men* who headed the divisions (these "officers" receive his sincere thanks at the end of his 1872 Report to the secretary of the interior). Perhaps this explains the wider pattern as well: the divisional structure of a large army was simply the only available organizational model during the early stages of bureaucratic growth, and the more modern functional division of labor appeared only haltingly and gradually.

In any event, Walker's organization of the central office highlights many of the issues involved in the politics of centralization. If it was to be done in a way that promoted a public image of objectivity and impartiality, centralization had to include not only better central control of hiring and the propagation of detailed standards of observation in the field, but also careful attention to who worked in the central office, where they came from and how they were hired. In contrast to his predecessor, Walker showed regional impartiality in hiring central office clerks, thereby doing as much as he could to neutralize the compromising effects of the still-necessary reliance on party patronage machines. Walker's commitment to a standard of competence by examination had as one of its uncomfortable side effects the impossibility of justifying the exclusion of women from the clerical workforce. But women in the central office conflicted with the prevalent cultural association between objectivity and maleness. Walker's response, at least at first, was to keep women carefully circumscribed in their own office spaces. Stepping back further, by associating the mobility and impartiality of enumerators in the field with maleness, and by organizing the central office around a quasi-military hierarchy of men (with women segregated into their own division), Walker invested the Census with the same gender valence given to public and scientific activities more generally by the male elite of the late nineteenth century.

The 1874 *Statistical Atlas* as a construction of the American social body

Through all of the efforts recounted above, to render the field of observation open and accessible to the governmental gaze, to fix the units of enumeration at known locations within it, to standardize acts of observation, to control the observers themselves, and then to capitalize on the perceived impartiality of observation by rendering compilation as efficient and accurate as possible, Walker pursued a comprehensive spatial politics of governmental knowledge. Its primary goal was to make the census as objective, as a-political, as possible, and, at least as important, to make it seem that way to the public. In so far as there was a "payoff" or culmination of this vast effort to achieve credibility, it came when Congress and the nation accepted the *compiled* results as a straightforward reflection of the reality of the American social body, despite the fact that compilation involved countless profound reductions, simplifications and selections of the incoming information. Edney's characterization of archives and museums applies here to the Census Office: "Data and artifacts can be collected within sturdy walls and there reassembled into meaningful arrangements . . . Within the museum, curators identified relationships between the samples themselves, rather than between each sample and its original environment, thus creating new complexes of knowledge in the form of artificial taxonomies."[85] If Walker's efforts to achieve abstraction, assortment, centralized control and transparent compilation were perceived as sufficiently successful, his rearrangements and artificial correlations of census data would appear utterly unpolitical in the eyes of the public.

The effects of the "license to compile" which Walker gave himself by way of his spatial politics of observation can be most clearly read in his 1874 *Statistical Atlas of the United States*. In 1875, the year following its publication, Walker offered the public a vivid image in which it implicitly plays the important role of a "screen" on which truth is projected: "The many millions of rays that fall confusedly upon the lens which is held up before the nation, are cast upon the screen in one broad, unbroken beam of light, truth pure, dispassionate, uncolored."[86] The title page of the *Atlas* encourages readers to assume that the compilation carried out by the "lens" is an utterly disinterested process, by trumpeting in bold letters the "contributions from many eminent Men of Science and several Departments of the Government" (Figure 2). The avowed purpose of the *Atlas,* according to the secretary of the interior, was to promote "that higher kind of political education which has hitherto been so greatly neglected in this country, but toward which the attention of the general public, as well as of instructors and students, is now being turned with the most lively interest." Its distribution was to "practically inaugurate the study of political and social statistics in the colleges and higher schools of the land."[87] I will close this chapter with a reading that attempts to ferret out the principles of selection and arrangement whereby Walker constructed the *Atlas*, and thereby suggests important features of the American social body he constructed to serve as the basis for the rest of his governmental program. The politics of compilation can be broken down into three different but related areas: the politics of topic selection, the politics of ordering or arrangement of material, and the politics of presentation. Together these establish emphasis and narrative structure, suggest causality more or less subtly, and construct the most important boundaries and constituent elements

---

[85] Edney, *Mapping an empire*, 39–40.   [86] Walker, "Our domestic service," 225.
[87] Quoted in F. Walker, "Preface and introduction to the statistical atlas," in F. Walker (comp.), *Statistical atlas of the United States* (Washington, DC, 1874), 1.

of the "geo-body" being represented.[88] I will first simply describe the contours of these three different kinds of politics, and close the chapter with a synthetic interpretation of their import.

The "Index to maps and charts" near the beginning of the *Atlas* divides the plates into three parts: "physical features of the United States," "population, social and industrial statistics" and "vital statistics." The first of these embraces topography, geology, coal deposits, river basins and climate features; the second, a hodgepodge of population density maps, a chart showing elements of population in each state, maps on illiteracy, wealth, debt and taxation, charts on "gainful employment" and religious affiliation, a fiscal chart of US government revenues and expenditures, maps of crop cultivation, and a page of maps covering different aspects of the Pacific Coast. The final part includes a map on "predominant sex," charts on age and sex characteristics of the living population in each state, a map of birth-rate variation, and a considerable number of charts on mortality, blindness, deaf-mutes and idiocy, each broken down by state into age–sex pyramids, and often according to race and nativity as well. All of this is followed at the end of the *Atlas* by a series of "Memoirs and discussions" of some of the topics treated in the maps and charts.

As noted in Chapter 2, the major "grids of specification" which structured the census included, of course, a geographical grid according to which the territory was composed of a set of regions defined by boundaries; a political grid which ensured that *individual states* would often appear as the most important regions, a grid of race and nativity, as well as grids of age and sex. These last three grids were often manifested as "secondary" variables against which "primary" ones (for example, population, or mortality) were broken down. These grids necessarily also provided much of the basic structure of the *Atlas*. Perhaps the most fundamental emphasis that results from their interaction is a reification of region and location as foundational variables. It is a short step from seeing that a phenomenon varies *with* place to assuming that it varies to some extent *because of* place. The other most prominent emphasis is on race and nativity groupings, which Walker terms the "principle elements of the population." Unlike age and sex, race and nativity appear *not only* as secondary variables in the charts on vital statistics *but also* as the primary subjects of some of the population maps in Part II. Nativity in particular is constructed by Walker out of whole cloth (Figures 3, 4 and 5). Despite the fact that it is an amalgam made up of peoples of British, Irish, French, German and Scandinavian origin, Walker insists on treating "foreign" as a biologically significant category, particularly in his essay at the end on "the relations of race and nationality to mortality in the United States." There he goes to great

---

[88] I borrow the term "geo-body" from Thongchai Winichakul's *Siam mapped: A history of the geo-body of a nation.*

lengths to explain how one would use statistics to establish scientifically the specific diseases to which "foreigners" are susceptible.[89]

Another important emphasis overlaid upon these basic ones is expressed in a visual pageant naturalizing manifest destiny (Plate XV on successive acquisitions of territory, followed by Plates XVIa, XVII, XVIIa and XIX, showing the westward movement of population from 1790 to 1870, with advancing population centroids marked) (Figure 6). The population maps Walker constructed featured blue lines to mark the boundaries of white settlement. These lines were located on the basis of calculations of white settlement density, and the Census Bureau's practice of marking these boundaries in the next two censuses would form the "factual" basis for Frederick Jackson Turner's influential frontier hypothesis. In essence, with these blue lines, Walker was delimiting the American social body at its westward extremities, and reinforcing the assumption that it was properly a *white* social body. As we shall see in Chapter 7, the seemingly obvious delimitation of the American social body at its eastward edge by the Atlantic Ocean was for Walker in fact much more problematic. Finally, among other topics notable in the *Statistical Atlas* for their unusual prominence are climate variability in the Part I; and an inflation of the significance of "afflicted classes" in proportion to their importance in the actual enumeration, mostly in Part III.

If we take as a baseline the questions on the census schedules, we can move from a consideration of the matters *emphasized* to a consideration of what was *de-emphasized* or selected out in the process of compilation. The emphases I have already noted will begin to make more sense overall once we see what Walker left out. The most glaring gap between the amount of information gathered by enumerators and the amount eventually represented in the *Atlas* is in the area of manufacturing and industry. Although Part II is entitled "population, social and industrial statistics," the last of these raises false expectations. Manufacturing and industry appear only in two figures, and then only indirectly. Plate XXXII is a chart consisting of a separate square figure for each state; rectangular sections within each square are given different colors to indicate the proportion of that state's population in each of a few broad classes of "gainful occupations or school attendance" (Figure 7). Manufacturing is one of these broad classes, but no more detail is given. The next figure (Plate XXXIII) is a national map of "wealth," but it is based on individual reported wealth, not on the manufacturing inquiries. In 1870 there was only one schedule for "industry," but it contained eighteen questions, more than a third as many as the entire population schedule. These questions covered such interesting matters as number of employees, amount of capital

[89] F. Walker, "The relations of race and nationality to mortality in the United States," in *Statistical atlas of the United States*, 2.

invested, value of products and wages paid.[90] The erasure of manufacturing goes hand-in-hand with an erasure of the cities in which it was largely taking place. Despite the fact that cities were the most dynamic places in the country in a variety of ways, Walker gave them almost no attention. He decided to exclude all towns of more than 8,000 residents from his calculations of population distributions for the corresponding maps. In one of the two "memorials and discussions" he contributed at the end of the *Atlas*, on "The progress of the nation, 1790–1870," he paused long enough merely to note that it is difficult to define a city.[91] There are no "close-up" maps of cities, and none of the information in the *Atlas* is differentiated according to an urban–rural distinction. Crime and pauperism, two phenomena which would have showed marked associations with both manufacturing and urbanization, are never depicted in any form. Another area of neglect is education. Despite the fact that the enumerators had ten questions to ask on education, including, for example, number and sex of teachers and students, it shows up in the *Atlas* only as "attended school," a category of undifferentiated adults listed alongside "gainful employment" in the chart discussed above (Figure 7). Although this is sheer speculation, it may be that Walker de-emphasized education because he was not anxious to make public the preponderance of women in the public role of educators.

The ordering or narrative arrangement of the material in the *Atlas* leads from physical/resource topics to population, then to the activities (broadly conceived) of the populace, and finally, to vital statistics. Lest there be any doubt about whether relations between the figures are legitimate objects of curiosity, Walker reassures the reader that the "highest use of these Maps and Charts is when they are compared with each other, so far as their subjects are cognate in any degree, for the discovery of relations and proportions which cannot be made to appear in any one map."[92] The comparisons he has in mind are made clear in the final paragraph of his "Preface and introduction," which bears quoting in its entirety:

No reference has been made to the use of the physical maps in Part I, in explaining the facts of vital, social, and industrial concern which are represented graphically in Parts II and III. To exhaust this subject would require a volume; only to open it, an extended article. The relations to animal life and health, to vegetable growth and reproduction, and even to industrial development, which are sustained by temperature and humidity, both in their mean and in their extreme range and variability, by the pressure of the atmosphere and the movement of the air, by the character of the soil, its drainage and the extent of its tree-covering; these relations do not so much as need to be alluded to

[90] Wright and Hunt, *History and growth*, 87, 107–108.
[91] F. Walker, "The progress of the nation, 1790–1870," in *Statistical atlas of the United States*, 5 (each essay is independently paginated).
[92] Walker, "Preface and introduction to the statistical atlas," 5.

here, in justification of the inclusion, in a statistical atlas, of maps illustrative of the physical features and meteorological conditions of the country. The compiler trusts, not only that this juxtaposition of the two orders of facts will afford the true explanation of a vast number of phenomena seeming most strange and contradictory of recognized causes in the political and moral constitution, but that an illustration, so large and varied, of the effects of physical influences upon the progress of population and the condition of society, may even serve to suggest to the physical geographer some possible modifications of his own generalizations.[93]

This strong geographical determinism explains the narrative structure of the *Atlas*: it presents a causal sequence, leading from physical enviroment to social phenomena. More particularly, it explains the maps of storm centers and temperature extremes (in line with the relatively common racist belief that Anglo-Saxon "superiority" had its source in the challenges posed by the bracing and variable climate of Northern Europe). Walker also considered elevation above sea level to be a significant factor in explaining racial distributions (thus the map on that subject); he also subscribed to a vaguely Jeffersonian–Turnerian view in which rural and frontier environments were most conducive to vigorous behavior and fertility of all sorts (thus the map of variation in birth-rates).

The third and final politics of compilation to consider before putting all of this together is the politics of presentation. I will do so only briefly, but even a rapid survey brings out important features. First, the preponderance of national maps, especially those showing exploitable resources and those charting the westward spread of the white population, encourage the assumption that the nation is a unified entity. Walker's habit of drawing a boundary around the white population in many of his maps, where whites per square mile dropped below two, gave a racial valence to that unity (Figure 6). It was also the foundation for Frederick Jackson Turner's famous claim in 1893 that the frontier had "closed."[94]

The *Statistical atlas* makes a very strong visual argument that the most significant building blocks out of which the nation is assembled are the individual states. Not only are state borders evident in the national maps, but almost every chart (save the Fiscal Chart of the US Government, Plate XXXVa) has as its basic components separate geometrical figures for each state. In general, the representation of gainful employment, school and religion, on the one hand, and most of the "afflicted classes" on the other (Figure 8), in charts rather than maps *reflects and reinforces the association of these phenomena with state governments*. While the statistics on religion, like those of crime and pauperism, were collected by correspondence with the heads of institutions, and thus could not be precisely mapped, statistics on the "afflicted classes" were

---

[93] *Ibid.*
[94] F. J. Turner, "The significance of the frontier in American history," in *Frontier and section: Selected essays of Frederick Jackson Turner*, ed., R. A. Billington (Englewood Cliffs, NJ, 1961), 62.

taken as part of the house-to-house count, and thus Walker could have mapped them had he so chosen. The heightened focus on the states, then, represents a decision of some kind.

The chart on "Principle constituent elements of the population of each state" (Plate XX), and the charts showing "church accommodation" and "gainful occupations or school attendance" (Plates XXXI and XXXII) represent each state with a square set off in the white field of the page from the other states; the squares are arranged in rows from left to right and from top to bottom (i.e., in normal "reading" order), alphabetically (Figures 9 and 7). So arrayed, the boxes strongly suggest components or building blocks waiting to be assembled. In most of the charts comprising the "vital statistics" section, each state is represented by an age-sex diagram, and here, too, they are arranged in rows read alphabetically from left to right and from top to bottom of the page (Figure 5). This ordering on the page is certainly an example of the "new complexes of knowledge in the form of artificial taxonomies" which Edney identifies as the typical product of compilation.[95] One of its effects is to reinforce the invisibility of cities, and, in effect, to insist that a city's significance is absorbed in that of its state.

Perhaps the most interesting aspect of this new ordering is the fact that in all charts except "Principle elements of the population . . .," and the last four charts on blindness, deaf-mutes, insane and idiots (in which each state's complement is represented as a scaled circle), the boxes or diagrams representing states are of uniform size (Figure 7). New York is exactly the same size as Rhode Island or West Virginia, despite the vast differences in their respective populations. One implication that could be drawn from this is that Walker wanted states to be considered *statistically* as equals, as a "Senate" rather than as a "House of Representatives." In so far as this kind of thinking actually did play a role in Walker's decision-making, it would not be surprising. In American history, the Senate has always been perceived as more "nationally" minded than the House, which has often been demonized as the more unruly arena for naked, particularistic interests. Walker definitely understood the significance of the census in nation-building terms, as an instrument for the clarification and furtherance of the general interest.

In the four pie-charts rounding out the graphical portion of the *Atlas* (Plates LI–LIV), and in the Pacific Coast maps (Plate XXXVIb), Walker aggregated states into basic regions (Northern, Southern, Western, Pacific Coast and Territories), thus offering an intermediate construction between the two more dominant levels of nation and individual state (Figure 8). But the groupings do not carry much rhetorical weight, and are clearly derivative of their component states.

Stepping back now and considering how Walker's *Statistical Atlas* shaped

---

[95] Edney, *Mapping an empire*, 40.

the American social body through his politics of compilation, it is possible to distill a few basic messages. First, America is essentially a nation of space, not place; it is a vast, varying landscape whose "punctuation marks" (cities) are not central to its constitution, but whose internal (state) boundaries are. These state boundaries distinguish different "sub-societies" within the United States, each with its distinct "signature" of demographic and social conditions. America is a landscape brimming with resources and quickly being overspread by people eager to put these resources to use through farming, lumbering and mining. These people making use of public space are chiefly men (since the *Atlas* only shows women indirectly in the map of birth-rates). Manufacturing and the problems that go with it are a relatively minor class of phenomena. The American climate is a vigorous, challenging one, and the various "elements" of population occupying the land give this climate a range of racial and national "constitutions" on which to have its effects.

In more analytical terms, the *Atlas* constructs an American exceptionalist vision. The historian of science Dorothy Ross was correct to place Walker near the beginning of a tradition of indigenous social science dedicated to finding ways of preserving America's uniqueness *vis-à-vis* Europe.[96] The exceptionalist vision styled the United States as the virtuous land of widespread opportunity, a nation managing to avoid the class struggle that had beset all other industrial societies, and showing its character as the "chosen" society in a variety of ways, including rapid population growth. Walker's version of this idyll shows definite Jeffersonian features, particularly the antipathy to cities and manufacturing, and faith in the powers of rural or frontier life. The exceptionalist vision interpreted the upheavals of the era as a crisis. Walker's response as encoded in the *Atlas* is part denial (of class struggle, women in public space and urbanization), and part engagement (with the issue of immigration). The *Atlas* shows Walker at an early and relatively optimistic moment, not yet confronted with a labor movement at full power, nor with the torrents of immigration that would shortly come, but already wary, and focused on immigration and race in particular. He was constructing a nation, but he was worried about the "elements" that would compose it.

In Chapter 6, I offer a characterization of the ideological or interpretive lenses through which Walker made his "normalizing judgments" about the American social body he had constructed. This is the second "moment" of the cycle of social control, where "is" is weighed against "ought" in order to decide "what is to be done" about the problems besetting the American social order. This moment cannot be disentangled completely from the first, "observational" moment, as previously held judgments help construct observation, and observation in turn shapes subsequent judgments. Yet the question of how the American social body is ordered and should be ordered is different

---

[96] D. Ross, *The origins of American social science* (New York, 1991).

from the question of how best to look at it. In Chapter 6, I will explore Walker's more openly and self-consciously held position within the realm of political economy. Thus the chapter will be concerned more with overt principles than with implicit strategies.

As with any person who has spent decades writing about his or her society, it is difficult to reduce all of Walker's thought to a single, utterly consistent "position," but the "American exceptionalism" diagnosed in my reading of his atlas is a good place to start. In constructing Walker's political economy, I will also pay special attention to geographical issues, particularly his thinking about the social and economic significance of mobility and fixity.

1 Francis Amasa Walker

STATISTICAL ATLAS

OF THE

UNITED STATES

BASED ON THE RESULTS OF THE

NINTH CENSUS 1870

WITH CONTRIBUTIONS FROM MANY EMINENT MEN OF SCIENCE
AND SEVERAL DEPARTMENTS OF THE GOVERNMENT.

COMPILED UNDER AUTHORITY OF CONGRESS

BY

FRANCIS A. WALKER, M.A.

SUPERINTENDENT OF THE 9ᵀᴴ CENSUS,

PROFESSOR OF POLITICAL ECONOMY AND HISTORY,
SHEFFIELD SCIENTIFIC SCHOOL OF YALE COLLEGE.

JULIUS BIEN, LITH.

1874

2 Title page, *Statistical atlas of the United States*, 1874

3 Colored population, 1870

4 Foreign population, 1870

5 Native and colored population by states and territories, 1870

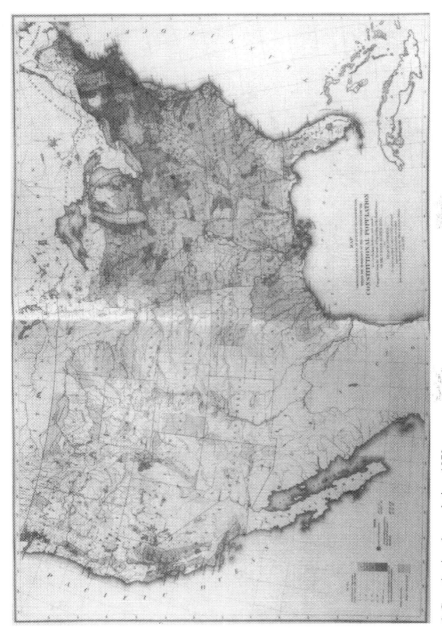

6 Constitutional population, 1870

7 Gainful occupations or attending school, 1870

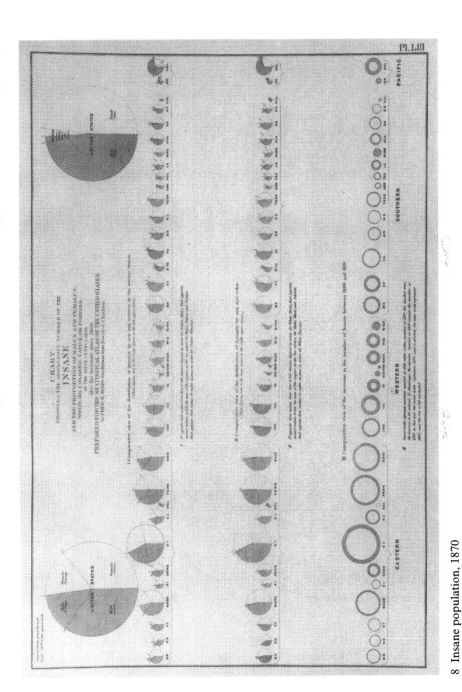

8 Insane population, 1870

9 Principal constituent elements of the population of each state, 1870

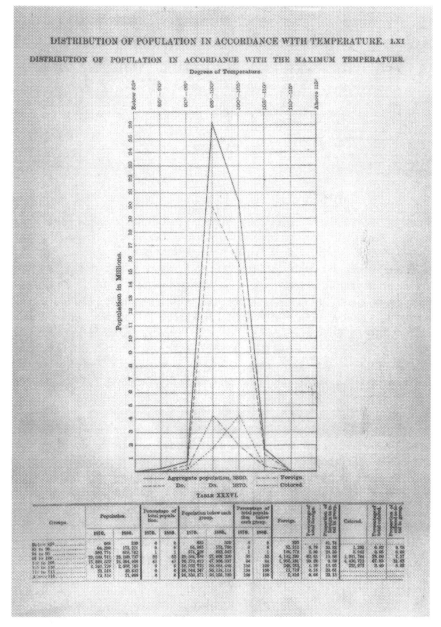

10 Distribution of the population in accordance with temperature, 1880

6

An American exceptionalist political economy

This chapter structures an outline of Walker's political economy around three main preoccupations associated with his American exceptionalism: a theory of distribution which could allow Walker to argue that class struggle was not necessary in America; a preoccupation with race and nativity as forces affecting the American political economy; and a deep concern with population growth rates as indicators of national health. After briefly situating Walker in the history of political economic thought, I will present the outlines of his political economy (with particular reference to his evolving theory of distribution), then explain in greater depth the ideas behind the prominent role he gives to race and nativity. Next, I will trace the way his growing concern with immigration altered his thinking on population growth, and at the end of the chapter briefly suggest that all of these concerns converge on the issue of *manhood*. While issues of class and race certainly shaped Walker's thought, their effects crystalized in the figure of the embattled American man, and it was to save the latter's manhood that Walker would go on to propose what he did in the way of social regulation (the subject of Chapter 7). In short, this chapter is about how the ordering principles already enshrined in his statistical construction of the American social body at the moment of observation converged and meshed with contemporary discourses of political economy, race and nationalism to produce a specific view of what was wrong with the country.

Western political economic thought was in a period of transition during the late nineteenth century. Haltingly and unevenly, a shift was taking place in the scope, methods and issues involved in the study of economic processes, and the replacement of the term "political economy" by "economics" was a telling symptom. As Blaug notes, it was anything but a clearly demarcated revolution, but all the same, there was a fundamental and systematic loss of interest in much of the classical problematic that had dominated up to that time.[1] The classical political economists (who took their points of departure from Adam

[1] M. Blaug, *Economic history and the history of economics* (New York, 1986), 209–218.

Smith, David Ricardo and Thomas Malthus) considered economic activity to consist of four basic "departments," each of which involved its own distinctive processes. The overarching question was, "how is wealth created and what happens to it once it is created?" Whatever their specific answers, most political economists presented them within the framework of the four departments: production, distribution, exchange and consumption.[2] Each of these four stages in "what happens to wealth" was assumed to operate according to its own dynamic. One shorthand way to understand the sea change that had occurred by the early decades of the twentieth century is as the takeover or colonization of other departments by that of exchange. Eric Roll puts it this way:

> The first thing which confronts the [typical modern] economic theorist is an economic reality which in spite of all its complexity is at once reducible to a network of exchange transactions in the market . . . [E]ven the transactions of the production process are seen to resolve themselves into the purchase and sale of raw materials, capital goods, money capital, and labour. If, then, we regard the economic system as an enormous conglomeration of interdependent markets, the central problem of economic inquiry becomes the explanation of the exchange process, or, more particularly, the explanation of the formation of price.[3]

Price, in turn, came to be thought of as determined less by the *cost* of producing a particular supply than by the *utility* of products from the perspective of demand. This shift manifested an abandonment of "an historical view of the structure of society which underlay the whole economic process," in favor of "a view of society as an agglomeration of individuals."[4] One of the logical results of this shift was that the buying and selling of labor lost its distinctive importance, and the category of class lost much of its previously central significance. According to Roll,

> it cannot be doubted that in its origins the utility school was often also influenced by a desire to strengthen the potentially apologetic aspects of economic theory. The classical theory was not strong enough to withstand the attacks of the growing working-class movement. The claim that a certain social structure – particularly when, as in the work of Ricardo, this was shown to contain severe conflicts of interest – should be regarded as the end of history could not be logically defended. Nor could existing conditions be made palatable simply by appeal to universal laws.[5]

Standard texts on the history of Western political economy all devote at least some space to Francis Walker, though not all would agree with economic historians Gary Walton and Ross Robertson that he was "the dean of American economists" at the time.[6] Walker is generally placed somewhere in the transition from classical to modern neo-classical economics. The authorities Walker

---

[2] E. Roll, *A history of economic thought*, 5th ed. (Boston, 1992); J. Schumpeter, *History of economic analysis* (New York, 1954).   [3] Roll, *A history of economic thought*, 338.
[4] *Ibid.*, 339.   [5] *Ibid.*, 340.
[6] G. Walton and R. Robertson, *History of the American economy*, 5th ed. (New York, 1983), 459.

resorted to most often in his systematic textbook included "classical" political economists – Ricardo, Mill, Nassau Senior and J.E. Cairnes – but also representatives of "modern economics," particularly W. Stanley Jevons and Walker's friend Alfred Marshall. Walker was aware, for example, of the move to place the laws of exchange at the basis of the other departments, but refused to go along with it. For him, a "science of exchanges" still begged the question, exchanges of what?[7] Robert Heilbroner's popularly written *The Worldly Philosophers* simply dubs Walker "a leading professional economist in the United States."[8] Joseph Schumpeter, in his magisterial *History of Economic Analysis*, fleshes out the picture considerably, commenting on the wide range of pursuits occupying Walker's energies. Schumpeter attributes some of Walker's reputation to his high visibility: "his own contributions to economic theory (residual claimant theory of wages, emphasis upon the role of the entrepeneur, criticism of the wage-fund theory) received perhaps more attention than they would have if made by a less prominent man."[9] Eric Roll credits Walker for the incisiveness of his critique of the wage fund theory (about which more below), but in the end implicitly echoes Schumpeter's judgment: "Walker worked in a number of fields in all of which he distinguished himself by a considerable energy and by the vigorous espousal of definite views."[10] Bernard Newton, author of the only book on Walker's economics, hails Walker as "the most prominent economist and statistician on the American scene during approximately the last quarter of the nineteenth century," and gives him a pivotal place in the larger conceptual transition to marginalism: "Walker's historical tasks were to challenge and lessen the impact of some of the 'classical' ideas, to stimulate a reconsideration of these ideas, and to develop new conceptions which helped lead American economics from its former course in the direction of an indigenous 'neoclassicism.'"[11]

Politically speaking, Walker's overall approach to industrial society places

[7] F. Walker, *Political economy*, 3rd ed. (New York, 1888 [1883]), 3.

[8] R. Heilbroner, *The worldly philosophers: The lives, times and ideas of the great economic thinkers*, 6th ed. (New York, 1986), 191.   [9] Schumpeter, *History of economic analysis*, 867.

[10] Roll, *A history of economic thought*, 385.

[11] B. Newton, *The economics of Francis Amasa Walker: American economics in transition* (New York, 1968), 1, 4. See also J. Dorfman, *The economic mind in American civilization*, vol. III, *1865–1918* (New York, 1949), 101–110. While Newton's is the only published volume on Walker's political economy, three unpublished graduate theses have treated it as well. John W. Kendrick, "The economics of Francis Walker," MA thesis, University of North Carolina (1939) concentrates on Walker's writings concerning money and bimetallism; Robert B. Ekelund, "Contributions of Francis Amasa Walker to economic thought," MA thesis, St. Mary's University, San Antonio, TX (1963) focuses on Walker's theories of distribution and critique of the "wage fund" theory; finally, Richard J. Sidwell, "The economic doctrines of Francis Amasa Walker: An interpretation," Ph.D. thesis, University of Utah (1972) treats those aspects of Walker's political economy most relevant to the social and political upheavals of late nineteenth-century industrialization (distribution theory and specific writings on social problems).

him somewhere between the poles of simplistic apology for capitalism (in line with the doctrine of *laissez-faire*) and strident advocacy of drastic change (*à la* Edward Bellamy, Henry George or Karl Marx). Although Walker was indisputably closer to the pro-business end of the spectrum, he was not a mere toady for the Gilded Age. Walker embraced the notion that free competition is beneficial in principle, but insisted that in actually existing historical conditions, making competition the only goal of economic policy would be sheer folly.[12] He considered private property the best foundation for any viable economy, but recognized that patterns of property ownership put most industrial workers at a real disadvantage in the struggle for wealth.[13] He believed that an increase in the money supply, favoring debtors at the expense of creditors, was a form of "confiscation," but he also felt that large corporations were injurious to fair competition.[14] In Walker's influential systematic text, *Political economy*, he neatly summarized the basic mixture of German and English traditions on which he had grounded his own approach: "Ricardian political economy . . . should constitute the skeleton of all economic reasoning; but upon this ghastly framework should be imposed the flesh and blood of an actual, vital political economy, which takes account of men and societies as they are."[15] Echoing J. E. Cairnes, Walker believed that political economy should seek to discover the most important laws of the production and distribution of wealth, but also to take account of the complicating human and natural contexts in which those laws must operate.[16] Walker saw much mischief arising from John Stuart Mill's artificially constructed "rational economic man": "Economists should be careful how they apply to mankind as they are, conclusions from the study of such a monstrous race, made up entire of laziness and greed, incapable of love or hate or shame."[17] I will present Walker's political economy in two parts, corresponding to his two major treatments of distribution (*Wages: A treatise on wages and the wages class*, published in 1876, and *Political Economy*, published in 1883). The key concepts through which Walker explains departures from Mill's "rational economic man" and other principles of abstract economics are, first, *mobility of labor*, and later, *labor productivity*. Both concepts are present in both major texts, but the emphasis changes markedly from the 1870s to the 1880s. Again,

[12] Newton, *The economics of Francis Amasa Walker*, 22–23, 165–167.

[13] F. Walker, "Private property," unpublished lecture, reprinted in *DES*, vol. II, 405–411; Newton, *The economics of Francis Amasa Walker*, 159.

[14] Walker, *Political economy*, 163; "Socialism," *Scribner's Magazine* 1 (1887), 107–119, reprinted in *DES*, vol. II, 264–266.

[15] Walker, *Political economy*, 17. For a compact review of Walker's departures from the *"a priori* school" of English political economy, see "The present standing of political economy," *Sunday Afternoon* 3 (1879), 432–441, reprinted in *DES*, vol. I, 301–318.

[16] Walker, *Political economy*, 12–13; F. Walker, "Cairnes' 'Political economy'," review of J. E. Cairnes, *Leading principles of political economy*, reprinted in *DES*, vol. I, 279–280.

[17] F. Walker, *Wages: A treatise on wages and the wages class* (New York, 1876), 366.

an emphasis on geographical aspects of Walker's thought will help to keep in view the territorial features of his governmental program.

Distribution and mobility

The first extended presentation of Walker's views was his *Wages*. Probably the most influential argument he made there was an argument against "wage fund theory," a transparently anti-labor piece of economic reasoning.[18] The wage fund theory, which was accepted by many American political economists after the Civil War, was essentially an attempt to make agitation for higher wages appear futile and pointless, by fixing as unalterable the size of the overall fund used to pay workers at any one time. Wage fund theorists claimed that wages were paid out of the total quantity of capital on hand (i.e. out of past earnings), and that since this was a fixed amount, any gain in earnings by one group of workers in one sector or industry would automatically deprive other workers of the equivalent amount somewhere else in the economy. At any one time, wage distribution was a "zero-sum game" among workers. By contrast, Walker, like Marx and others, saw current production and the future revenues it would generate as the source of wages. The employer might pay workers out of current capital, but the amount paid would be based on expectations of future revenues.[19] Thus, if productivity is growing, it is not automatically futile to press for higher wages. Unlike Marx, however, Walker credited both employers and workers with the production of value, the former through the exercise of "technical skill, commercial knowledge, and powers of administration" and the latter through the material exercise of their labor power.[20] Because he had two classes participating in value production, Walker could not develop a clear means by which to compare the cost of labor with the value it generates, and so could not (even had he wanted to) produce a precise theory of exploitation.[21] However, he was very much alive to the possibility that employers could claim inordinate profits at the expense of workers' wages.

In *Wages*, Walker identified three major destinations for distributed wealth: capital (which in this early treatise included rent, and would bring a fixed return of interest for the use of wealth or land), employers' profits, and workers' wages.[22] Walker insisted that neither employers nor workers could exist without the other, and thus that the conflict of interest between them was

---

[18] Ekelund, "Contributions of Francis Amasa Walker to economic thought," 107.

[19] Walker, *Wages*, 131–133, 144.   [20] *Ibid.*, 245.

[21] Newton, *The economics of Francis Amasa Walker*, 93–96.

[22] The distinction between the role of the capitalist and that of the employer was the main contribution Walker's father Amasa Walker made to American political economy. The elder Walker had taken up the study and teaching of political economy after leaving the business world in middle age. See Newton, *The economics of Francis Amasa Walker*, 6; F. Walker, "The source of business profits," *Quarterly Journal of Economics* 1 (1887), 265–288, reprinted in *DES*, vol. I, 362, note.

fundamentally circumscribed. However, in this early formulation at least, something like exploitation could arise if a deliberate or accidental underestimation of productivity when setting wages allowed an excessive remainder of revenues to go into employers' pockets. Other conditions, discussed in more detail below, usually kept wages lower than they would otherwise need to be, but for Walker, this did not alter the fact that the idea of an *a priori* ceiling on wages was a myth.

Many of Walker's contemporaries would have agreed with his rejection of the wage fund theory, but some would have countered that a capitalist economy is self-correcting: excessive profits allow an expansion of productive investment and hence an increasing demand for labor which drives wages up. Alternatively, low wages could allow lower prices, lowering the cost of living and thus maintaining the buying power of wages at a constant level despite their numerical decline. Walker demurred from this view, pointing out, that (a) excess profits are not automatically reinvested, but may be dissipated in luxury spending; (b) partly for this reason, lower wages do not necessarily translate into lower prices for the worker *as consumer*; and (c) sufficiently low wages can easily render workers powerless to "compete" effectively with employers in the future, can increase their dependency and perpetuate their disadvantage.

With less food, which is the fuel of the human machine, less force would be generated; with less clothing, more force would be wasted by cold; with scantier and meaner quarters, fouler air and diminished access to light would prevent the food from being fully digested in the stomach and the blood from being fully oxidized in the lungs, would lower the general tone of the system and expose the subject increasingly to the ravages of disease. In all these ways, the laborer would become less efficient, simply through the reduction of his wages.[23]

In sum, the economic "harmonies" which many supposed would operate to restore a natural equilibrium distribution of wealth were for Walker a fiction.[24] The peroration that Walker places at the heart of the concluding chapter of *Wages* is scarcely that of a pro-capitalist ideologue. In reviewing what he considers to be the most crucial claims made in the book, the point on which he "most strongly insists" is "the doctrine that *if the wage laborer does not pursue his interest, he loses his interest*" (emphasis in original).[25] For Walker, the policy of *laissez-faire* corresponding to the theory of the harmonies, "like fire or water, is a good servant but a bad master."[26] In other words, Walker followed a middle path. He saw laborers and capitalists in terms that might describe two sports teams, competing vigorously, but within a structured context which benefited them both, and without which neither would prosper. The maintenance of this structured context justified certain regulatory interventions, as we shall see in Chapter 7.

---

[23] Walker, *Wages*, 287.   [24] *Ibid.*, 158–164.   [25] *Ibid.*, 411.   [26] *Ibid.*, 413.

[T]he distribution of the product of industry involves what may be termed a perpetual contest between the parties to production. This contest is not a destructive one, since the interest of each of the participants requires the existence, and, by consequence, the sustentation, of all the others. Yet, within the limits consistent with this, there is opposition of interests.[27]

Competition between the two classes would benefit both:

This . . . is the ideal industrial condition: that the body of laborers shall be able to offer an adequate economic resistance to continuous pressure from the employing class, so that no favors need be asked, on the one side, so that there need be no flinching, on the other, in the exaction of all which the most vigorous prosecution of self-interest may require.[28]

Whatever this was, it was not full-blown class struggle, with one class calling into question the legitimacy of the capitalist economy as a whole and the other reciprocating by abandoning any concern for the welfare of its opponent. Nor was it the political-economic embodiment of social Darwinism. In a formal sense, competition between economic actors was, like competition between animal or plant species, considered by Walker and others to be "natural," and considered therefore to produce "natural" outcomes.[29] But political economists of the *laissez-faire* variety could not consistently embrace social Darwinism and at the same time profess a doctrine of "harmonies" or self-adjusting economic forces. The "harmonies" implied that there was a reliable "floor" under wage levels, below which wages could not drop; by contrast, social Darwinism implied no such floor. The losers could in principle continue to decline until they perished. Whatever "survival of the fittest" entailed, it did not so neatly underwrite the status quo; it was a more dynamic doctrine, one which did not presuppose that everyone would benefit from competition.[30] Compared to the strict *laissez-faire* people, then, Walker might have appeared to tend toward social Darwinism or some variant. But his belief in the importance of competition, and his skepticism about the claim that the capitalist system *automatically* protected the interest of labor, were tempered by the realization that true competition was rare if not unknown. The structural preconditions necessary for a form of competition in which the losers could be abandoned to their fates in good conscience would have to be deliberately constructed. Rather than ignore all this and simply insist that the abstract principle of competition was the only basis for political economy, Walker attempted to give both elements, principle and fact, equal weight. He was neither a social Darwinist nor the opposite.[31]

---

[27] Walker, *Political economy*, 191–192.

[28] F. Walker, "What shall we tell the working classes?," *Scribner's Magazine* 2 (1887), 619–627, reprinted in *DES*, vol. II, 309.

[29] Ekelund, "Contributions of Francis Amasa Walker to economic thought," 28–31.

[30] R. Hofstadter, *Social Darwinism in American thought* (Boston, 1944), 143–145.

[31] See Sidwell, "The economic doctrines of Francis Amasa Walker," 125–129, for an alternative interpretation of Walker's relation to social Darwinism.

Walker wanted to bring such principles down to earth by considering all the concrete factors which operated in actually existing societies to modify, thwart or transform economic laws. Throughout his texts on political economy, he maintained a consistent focus on conditions that tended to prevent workers from pursuing their interests to the highest possible degree. In *Wages*, the paramount issue was worker mobility; in *Political economy*, it became productivity. But both issues are present in both texts, and Walker never abandons his views on either aspect of labor welfare. It was more a matter of changed emphasis than changed theory.

In the earlier text, Walker insists that it is *mobility* (both geographical and occupational) above all else which equips labor for the healthy sort of struggle with employers. In the struggle for a better situation, workers must have the ability always to search out the most advantageous job opportunities wherever they arise.

The laborer's practical ability to seek his best interest is made up of a material element – the means of transportation and present subsistence – and of intellectual and moral elements quite as essential, the knowledge of the comparative advantages of different occupations and locations offering themselves, and the courage to break away from place and custom to seek his fortune elsewhere.[32]

Mobility is necessitated by frequent large-scale changes in the landscape of industrial capitalism. In his later text, Walker used vivid imagery to illustrate his point:

Deal the heaviest blow you can with a hammer into a bin of barley, and you will not injure a single grain, though the hammer be buried to your hand, because every grain moves freely from its place, and the mass simply opens up to receive the intruding substance and closes around and above it.[33]

The blows that can befall labor in any particular locality can come from technological advances and from externally determined shifts in the local mix of occupations.[34] The latter danger is exacerbated "when we consider that, in the development of modern industry, trades become highly localized, entire towns and cities being given up to a single branch of manufacture."[35] Walker illustrates this danger by noting the devastation wrought in Yorkshire when the hoop skirt went out of fashion and put out of work the many factories there producing crinoline wire.[36] Contrary to the *laissez-faire* doctrine of the economic harmonies, the disadvantage wrought by immobility could become permanent, as workers not able to change location or occupation can easily sink into a self-reinforcing hopelessness and squalor.[37]

At a more local scale, geographical mobility is also crucial. Walker quotes his contemporary Frederick Harrison to the effect that the laborer, in order to pursue his best interest, "must himself be present at every market, which

---

[32] Walker, *Wages*, 338. [33] Walker, *Political economy*, 264. [34] Walker, *Wages*, 188–189.
[35] *Ibid.*, 203. [36] *Ibid.*, 179, note. [37] Walker, *Political economy*, 73, 265.

means costly, personal locomotion. He cannot correspond with his employer; he cannot send a sample of his strength; nor do employers knock at his cottage door."[38] In the midst of a long explanation of why women are less suited to industrial labor than men (about which more below), Walker paints a vivid picture of the travails involved in finding work:

> It must be remembered that it is not a question merely of taking a journey from home to a place where a "situation" has already been engaged, but, it may be, of seeking out employment from street to street, and from shop to shop, by repeated inquiries, and often through much urgency of application. This is what men have to do to "get a place," often going to doubtful localities, freely encountering strangers, and sleeping in casual company.[39]

This passage is clearly intended to evoke moral outrage at the prospect of women competing for work, but it is nevertheless not an inaccurate depiction of the plight of many unskilled laborers in the late nineteenth century.[40]

Mobility is also important for workers in their role as consumers. Walker makes the basic distinction between "real" and "nominal" wages central to his discussion of wages, insisting that the former, comprising the "comforts, necessities and luxuries of life" are of greater significance in determining the adequacy of wages than is the quantity of money received.[41] Given this distinction, a variety of factors can influence how efficiently and wisely the worker transforms nominal into real wages through consumption.[42] In his explanation of the disadvantages to labor of a "fictitious currency" (i.e., one not backed by a quantity of precious metals), Walker again shows a special sensitivity to spatial issues. A fictitious currency allows prices to fluctuate more frequently and drastically, rendering it far more difficult for the consumer to decide whether a particular rise in prices is fair. Where with a stable currency, consumers might only need to compare prices at other shops occasionally, under a fictitious currency, remaining up-to-date on prices (and thereby ensuring the beneficial effects of competition between retailers) requires an excessive amount of traveling around.[43] The fact that many retailers enjoyed local monopolies made the need for mobility all the more pressing.[44]

The various kinds of geographical mobility discussed in *Wages* all require both time and money. Workers could quite easily end up trapped by long and exhausting work hours, as well as inadequate wages, in situations representing anything but their "best interests." Although "it is not necessary that the whole body of laborers should be organized like a Tartar tribe, packed and saddled ready for flight," a significant proportion do need to be relatively free to move.[45] To some extent, in so far as a worker's local region was home to a variety of industries, occupational mobility could substitute for large-scale

---

[38] Quoted in Walker, *Wages*, 184.   [39] *Ibid.*, 376.
[40] See D. Montgomery, *The fall of the house of labor* (New York, 1987), ch. 2.
[41] Walker, *Wages*, 12.   [42] *Ibid.*, 13.   [43] *Ibid.*, 312.   [44] *Ibid.*, 314.   [45] *Ibid.*, 180.

geographical mobility.[46] But all too often, occupational mobility was a side-effect of the lack of special skills, and was more than offset by the relative precariousness of the work available to unskilled labor.[47] Although Walker was never in sympathy with the radical positions available in his time, and despite his conservative views on gender and race, his understanding of the obstacles to workers' mobility led him to some relatively moderate policy positions (see Chapter 7).

In Walker's view, some obstacles to mobility were "natural" features of human existence, or "natural" characteristics of specific groups. The ineluctable immobilizing traits common to all humans are brought out by contrasting "men" with "merchandise." With respect to merchandise, "destitute alike of sympathies and antipathies, competition is so far perfect that it may be reasoned upon as if no obstruction to exchange existed. The one additional penny of profit will send the bale of goods to East or West, North or South, to kinsman or to stranger, to black man or white, with absolute indifference."[48] People, on the other hand, are all to some degree tied down by such factors as "mental inertia, love of country, love of home, love of friends."[49]

It might seem inconsistent of Walker to take such a dim view of mobility in his dealings with the spatial politics of census taking (see Chapter 5), while at the same time championing mobility as the *sine qua non* of a healthy workforce in *Wages*. But Walker took the "natural" limits to human mobility quite seriously, and was perfectly comfortable with their continued operation. The fact that he saw mobility as an industrial advantage for workers did not mean that he thought they should have as much of it as physically possible. The maximum mobility would in fact entail unacceptable moral costs. What he took to be the proper degree of mobility can be inferred from passages in both his earlier and later political economy texts. In *Wages*, Walker anticipates the question of whether "tramps" can be thought of as a healthy economic phenomenon since they are so mobile:

[T]here is no more virtue to relieve the pressure upon honest, self-respecting labor in the forces which direct the movement of the "tramp," than there is of virtue to save men from drowning in the forces which bring a human body to the surface after a certain period of putrefaction. The body comes up, indeed, but only when swollen and discolored by the processes of corruption; and so the laborer, who has lost his hopefulness and self-respect and become industrially degraded, whether by bad habits for which he is primarily in fault, or by the force of causes he had no strength to resist, wanders about the country begging his food and stealing his lodgings as he can; but his freedom, thus obtained by being loosed from all ties to social and domestic life, does not so much relieve labor as it curses the whole community, rich and poor alike.[50]

---

[46] *Ibid.*, 188–189.   [47] See Montgomery, *The fall of the house of labor*, ch. 2.
[48] Walker, *Wages*, 163.
[49] *Ibid.*, 164. Here, Walker's formulation is strongly shaped by Cairnes; see Walker, "Cairnes' 'Political Economy'," 281–282.   [50] Walker, *Wages*, 204–205.

These were undoubtedly some of the people he had in mind in exhorting his census enumerators to scour all possible hideaways for members of the "floating population." The condition of being "loosed from all ties to social and domestic life" was unacceptable. This concern was also behind Walker's reluctance to support women who wished to pursue waged work outside the home. The costs of such work include:

Ill effects on the health of the family, on the duration of the [male] laboring power, and on the moral elements of industry . . . Waste in food, clothing and utensils; waste in laboring-force through ill-prepared and ill-preserved food; waste of the vital endowment of the rising generation through lack of that constant care which is the essential condition of well-being in childhood; waste of character and the formation of indolent and vicious habits through neglect to instruct and train the young, and through making the house cheerless and distasteful to the mature.[51]

At the other end of the spectrum, Walker believed it possible for laborers to be too tied down by family. The laboring classes could defeat their own interests by having too many children; the effects were as follows:

first, the reduction of vital force and labor power [attendant on too-frequent sex . . . see Chapter 4]; secondly, the diminution, perhaps the disappearance, of the subsistence fund heretofore laid up against the occurrence of bad seasons or the disability of the head of the family through accident or sickness, thirdly, the generation of infirmities and diseases of a transmissible character [a result of the overcrowding of living space].[52]

The golden mean for Walker was clearly the nuclear family, and he was willing to live with its immobilizing effects: "It is the solidarity of the family which prevents the law of the survival of the fittest from exerting that power in raising the standard of size and strength and functional vigor among men, which it exerts throughout the vegetable and animal kingdoms, generally."[53] In Walker's treatment of consumption in *Political Economy*, he naturalizes the isolated individual man as the basic economic unit, and then the nuclear family as the first set of extra responsibilities the individual man takes on as his income rises above the level of mere subsistence (the sequence of sections in the text run as follows: "§383. Subsistence; . . . §384. Clothing and shelter; . . . §385 [untitled]; . . . §386. The wife; . . . §387. The child; . . . §388. Children in excess . . .")[54] Although he probably believed that a nuclear family, like ownership of land, gives the individual man sharp incentive to be productive, Walker did not dwell on this point. In general, though, in the political economic as well as in the epistemological moment of Walker's governmental program, the compact nuclear family represented the proper compromise between fixity and mobility.

More easily changeable human laws and practices might also interfere with the mobility of labor. Under this heading, Walker briefly mentions fictitious

---

[51] *Ibid.*, 382.   [52] Walker, *Political economy*, 299.   [53] *Ibid.*, 300.   [54] *Ibid.*, 294–298.

currency and taxation of incomes, and spends quite a bit more time on "truck" (barter). Of more relevance here is his discussion of poor laws and parochial settlement laws in eighteenth and nineteenth-century England. English poor laws, by encouraging dependency, "brought the working classes to the verge of ruin, creating a vast body of pauperism which has become hereditary, and engendered vices in the whole labor system of the Kingdom which work their evil to this day."[55] Walker could hardly muster sufficient spleen to express his contempt for parochial settlement laws, which prevented peasants from moving out of the parishes in which they were registered:

[I]t is doubtful if all the barbarous enactments we have cited are together responsible for more of the present pauperism and destitution of England than is the Law of Parochial Settlement . . . It is only within the last twelve years that the cords that crossed the political body in all directions, cutting off the circulaton until every portion of the surface broke out in putrefying sores, have been loosened.[56]

Under parochial settlement laws,

all local calamities fell with unbroken force upon a population that had no escape . . . Industry might look up again, but the peasant, broken in his self-respect, brutalized, pauperized, could never afterwards be the same man. Employment might revive, but no art of man, no power of government could reconstitute the shattered manhood.[57]

Finally, within the general run of humanity, Walker singles out a few *groups* who suffer from an extra increment of inherent immobility. The American workforce is for various reasons unusually mobile, but even within the United States,

[t]he exceptions to this readiness to follow industry in its movements are found among three classes: the newly emancipated slaves of the South, in respect to whom no explanation is required, that portion of our women who are compelled to enter the general market for labor, and, lastly, our foreign population . . .[58]

Women, whose presence in public life disturbed Walker, are immobilized by "physical weakness, by timidity, and by those liabilities to misconstruction, insult and outrage which arise out of their sexual characteristics." These characteristics together render them unfit for the rude and strenuous business of chasing jobs.[59] I will treat race and nativity as they impair mobility at greater length later in this chapter, but it is worth noting once again the association Walker draws between "inferior" races and nationalities, on the one hand, and femininity, on the other.

Distribution and productivity

Walker systematized his political economy in his widely used 1883 textbook *Political Economy* (reprinted in 1888), and in the process altered its emphasis,

---

[55] Walker, *Wages*, 319.   [56] *Ibid.*, 308.   [57] *Ibid.*, 309.   [58] *Ibid.*, 181.   [59] *Ibid.*, 376.

or rather, brought to the fore another "layer" of concerns. The shift saw the issue of labor mobility so prominent in *Wages* deemphasized in favor of a concern with *productivity*.[60] In effect, while Walker still acknowledged the importance of mobility, it had become a background issue. The laborer formerly hunting through cities and larger regions for the best possible situation has, by the later text, somehow found a job, and the focus is now much more on his (for Walker, the laborer is a man) behavior in the shop. If his wages are low, what is it about his work habits, his intelligence, etc., that explains it?

In *Political Economy*, Walker formalized a "residual claimant theory of wages." In *Wages*, he had not yet been clear about the *order* in which the shares of wealth going to different classes were deducted from the product. His first concern had been merely to establish that profits and wages were not in competition for a pre-established sum of wealth, but could *both* be enhanced by vigorous competition within and between the employing and the laboring classes. In the new rendering, by contrast, he was explicit: wages were in principle the sum left over after rent, interest and profits were deducted from revenues. Walker believed that the first three deductions were relatively fixed, and thus that workers would be able to claim any increase in revenue due to increased productivity.[61] In *Political economy,* also, he tried to be clearer about how class struggle was avoided. He argued, first, that the employer was indispensable, that "the armies of industry can no more be raised, equipped, held together, moved and engaged, without their commanders, than can the armies of war."[62] "Jealousy" of high profits on the part of laborers was for Walker misguided; workers should instead direct their ire at the incompetent businessmen whose high costs of production raised the prices of the goods they had to buy as consumers.[63] The profits going to employers should be understood as analogous to rent. They form no part of the price of a product, but are the returns to employers whose talents are above those of the average businessman, much as rent is a measure of the differential productivity of land over and above the poorest "no-rent" land at the margin of cultivation.[64] "That profits are not obtained by deduction from wages is equally clear, when we consider that the most successful employers pay as high wages as the employers who realize no profits."[65] The "residual" character of wages made it difficult to imagine how anything like exploitation could occur. In his enthusiasm to deny the necessity of class struggle, Walker claimed that,

---

[60] Walker, *Political economy*, 258–261; Roll, *A history of economic thought*, 385.

[61] Walker, *Political economy*, 249–258.

[62] *Ibid.*, 75. Note the parallel between this picture of the "captain of industry" and Walker's own attempts to live up to a military manhood ideal of generalship in his public positions (see Chapters 4 and 5 above).   [63] *Ibid.*, 240 241.   [64] *Ibid.*, 235–240.

[65] Walker, "The source of business profits," 370. It did not occur to Walker that this could be seen in a very different light, that one could reasonably ask why wages remain the same even when profits rise.

notwithstanding the formal attitude of the laboring class in industry, as hired by the entrepeneur class and working for stipulated wages, the normal operation of the laws of exchange is to make the former, in effect, the owners of the entire product, subject to the requirement of paying the definite sums charged against the product on the three several accounts of rent, interest and profits.[66]

Bernard Newton notes that Walker's "strong emphasis on the functional relationship between productivity and wages influenced the succeeding generation of economists to recognize that wages were at least partially dependent upon the level of productivity of the workers . . ."[67] Paul McNulty, an historian of labor economics, concurs, explaining in a bit more detail how the "residual claimant theory" earned Walker a pivotal place as a precursor of the "marginal productivity" approach to wages. By raising the issue of the variable productivity of labor, Walker opened up the possibility (pursued systematically by others) that variable wages could be explained as a result of market demand for a variably productive labor supply.[68] Indeed, Newton points out that it was in the midst of debate amongst American professional economists over Walker's residual claimant theory that John Bates Clark first applied marginal productivity theory to wages.[69] Politically speaking, the theory of distribution in *Political Economy* placed the burden of achieving adequate wages more squarely on the shoulders of *workers themselves* than had the earlier version, which had stressed structural factors more strongly. Thus the later theory can be seen as a modest move in the conservative, pro-business direction. Again, this shift was important in that it amounted to placing the workings of class and other "structural" factors "backstage," and shining the spotlight instead on supposedly inherent racial traits of groups of workers.

With productivity, as with mobility, Walker was concerned with the concrete conditions that affect it, not with abstract principles alone. Productivity was the lynchpin of Walker's residual claimant theory of the distribution of wealth. By the productivity of workers, Walker simply meant "their energy in work, their economy in the use of materials, or their care in dealing with the finished product"; the more productive worker was "more careful and painstaking, more adroit and alert, more observant and dexterous."[70] Such workers brought not only a reduction of waste but also an improvement in the quality of the products they made.[71]

The obstacles to productivity, like the obstacles to mobility, cover a wide range of conditions that are, strictly speaking, extra-economic. The difference

---

[66] Walker, *Political economy*, 258. Walker attributes this idea to W. Stanley Jevons; see Walker, "The source of business profits," 370.

[67] Newton, *The economics of Francis Amasa Walker*, 82–83.

[68] P. McNulty, *The origins and development of labor economics* (Cambridge, MA, 1980), 116; see Walker, *Political economy*, 94–101 for the discussion of what he called "final utility."

[69] Newton, *The economics of Francis Amasa Walker*, 92.

[70] Walker, *Political economy*, 251, 254.   [71] *Ibid.*, 255–256.

is that in *Political Economy,* Walker stresses those factors affecting the inherent qualities of individual workers rather than those residing in external economic and geographical conditions. In *Wages,* Walker already identified six broad categories of impediments to productivity: (1) "peculiarities of stock and breeding"; (2) "meagreness or liberality of diet"; (3) "habits, voluntary or involuntary, respecting cleanliness of person, and purity of air and water"; (4) "the general intelligence of the laborer"; (5) "technical education and industrial environment"; (6) "cheerfulness and hopefulness in labor, growing out of self-respect and social ambition, and the laborer's interest in the results of his work."[72] In *Political economy,* substantially the same factors are enumerated, and Walker adds the effects of a more or less well-developed division of labor.[73]

For Walker, race, nativity and manhood lie at the root of variation in most of these elements of productivity, as the following passage implies:

> The difference between an English woodsawyer, before a pile of hickory cordwood, and an effeminate East Indian, accustomed to think it a day's job to saw off a few lengths of bamboo, is not so great as that which would exist between a Maine mast-man and a Bengalee at the foot of a 40-inch pine. The one would lay the monster low in half a day, the other might peck at it a week and scarcely get through the bark.[74]

Although this preoccupation springs up throughout both of his major texts, there is much less to counterbalance it in *Political Economy.* Race and nativity are central issues for Walker, key factors which account for most of the differences between the behavior predicted by *a priori* economic principles and the infinite variety of actually existing economies: "As by [the *a priori*] view of political economy all men are taken as equally absorbed in the passion for wealth, so all men are taken as equally lazy and self-indulgent. The South Sea Islander and the large-brained European are equally averse to exertion; equally subject to the impulses of immediate appetite."[75]

Taken together, the many passages like these in Walker's writings reveal what might be called a moral dimension of his distinction between economic principles and economic facts. He is not in the end satisfied merely to point out that all societies depart from the ideals spelled out in *a priori* economic principles. Since departures from the ideal functioning of economic laws are usually traceable to the effects of inferior racial and national groups, these departures take on a negative coloring. Indeed, he sees the "native American race" as just the sort of mobile, vigorous, productive race best suited to withstand the unfettered operation of the laws of competition, if left to themselves. The difference between ideal economic behavior and actual economies varies with race and nationality, and "ideal" takes on the meaning of a desirable norm or goal, not merely a neutral, counterfactual abstraction.

[72] Walker, *Wages,* 49.   [73] Walker, *Political economy,* 46–56; 57–60.   [74] Walker, *Wages,* 42.
[75] Walker, *Political economy,* 15–16.

This moralization of Walker's economic principles had its greatest effect in his thinking about immigration from Europe. During the 1870s, Walker had believed that immigration could have beneficial economic effects, especially in those Western and Southern regions where neither the cultivation of land nor fledgling industries had reached the point of diminishing returns on additional labor. The counties in these regions clamoring for immigrants "are not acting foolishly. They are not calling in additional laborers to divide with them a pre-determined product."[76] Indeed, both his earlier and his later theories of distribution can be seen as relatively friendly to immigration, since they reject the notion of a pre-set wage pool and thereby deflate arguments that immigrants simply represent extra competition for the same wealth. However, Barbara Solomon's claim that in the mid-1870s Walker "accepted immigration as an unquestionable material good" needs some qualification.[77] Even as early as 1874, he was provoked by what he perceived as European arrogance into announcing that immigration from that continent "has exerted an influence prejudicial to the development of the higher and finer manufactures."[78]

Over the following ten to fifteen years, particularly after the "labor troubles" of the mid 1880s, Walker came to believe that immigrants were responsible for practically everything that was wrong with the otherwise healthy American political economy. It was immigrants specifically, not other "inferior races," whom Walker placed at the root of many of the obstacles to productivity discussed above. These immigrants had brought with them "lower standards of work and lower social ambitions, with less, at once, of general intelligence and of technical skill."[79] They explained as well most of the radical agitation for more serious class struggle, and were to blame for most incidents of violence such as the 1886 Haymarket bombing in Chicago.[80] Walker followed his animus against immigrants to its logical conclusion, insisting finally that it was immigrants who first introduced not merely class struggle, but *classes themselves* into the American social body.[81] All manifestations of class struggle in the European sense (unions, etc.) were for Walker signs that immigrants could not stand up to competition with employers *like true men*. Class divisions and true manhood were ultimately antithetical (see Chapter 4). Walker's growing disapproval of immigration was quite typical of the views generally held by the cultural elites of northeastern cities in everything but its intensity. Because his animus was so intense, and because he could marshal the cultural authority of statistical social science, he took a leading

---

[76] Walker, *Wages*, 149.   [77] B. Solomon, *Ancestors and immigrants* (Cambridge, MA, 1956), 71.

[78] F. Walker, "Occupations and mortality of our foreign population, 1870," from *Chicago Advance* Nov. 12, 1874, Dec. 10, 1874, Jan. 14, 1875, reprinted in *DES*, vol. II, 217.

[79] Walker, "What shall we tell the working classes?," 308.   [80] *Ibid.*, 312–313.

[81] F. Walker, "Methods of restricting immigration," *Yale Review* 1 (1892), 138–143, reprinted in *DES*, vol. II, 441.

position among his peers.[82] But, as with the uneasiness over women in public space and the threats posed by class struggle and urbanization, Walker's basic views were interchangeable with the views of hundreds of others in similar positions of cultural preeminence.[83] Again, his special role consisted not so much in what he thought about these things, but in the unique *web of discursive positionings* which allowed him to link together (and thus mutually reinforce) the authority of discourses concerning social statistics, political economy and the role of the modern state.

In short, race and nativity came to explain the gravest threats to Walker's American exceptionalism. Were it not for immigration, the United States would be much better able to avoid class struggle by fostering a community of productive men. For this reason it is worth devoting some space to a more extended discussion of Walker's views on race and nativity, and how his position related to prevailing thought at the time. The next section will touch on census results as they participated in Walker's reification of racial categories; environment and geographical determinism more generally, neo-Lamarckian racial thought, and various other ingredients in Walker's racial "brew." My goal will not be to render an exhaustive account of every thread woven into Walker's racial thinking, but rather simply to make some of the most salient features of his thinking intelligible. In the final section of this chapter, I will chronicle Walker's growing alarm at the relationship he saw between immigration and population growth rates. Immigration turned out to pose the most fundamental danger not only to America's *productive vigor* but also to its *reproductive vigor.* The figure at the intersection of these two concerns was the American man, and it was this figure Walker had in mind in proceeding from his political economy to his governmental policy recommendations (see Chapter 7).

Race, nativity and environment in Walker's thought

Walker drew on a number of strands of racial thinking current in his day, but made no explicit attempt to knit them into a unified racial theory. In a sense, his racial ideas merely constituted a sampling of the theories then circulating among the educated, northeastern elites discussed in Chapter 3. His position was distinguished from those of many contemporaries mainly by the tenacity and breadth of resources with which he pursued his indictment of European immigration. For Walker, as for many of his contemporaries, "race" could mean what it came to mean in the twentieth century, but it could also mean what would later be called "nationality" or "ethnicity," as well as nativity.[84]

[82] See Solomon, *Ancestors and immigrants,* for a still-fresh account of the anti-immigrant sentiments of the Boston elites.   [83] *Ibid.*

[84] D. Livingstone, *Nathaniel Southgate Shaler and the culture of American science* (Tuscaloosa, AL, 1987), 131.

Blacks constituted a race, but so did the French, and so did "foreigners" more generally.

In practice, Walker assumed a hierarchy of races that was vague at the lower end but more clearly ordered toward the top. Indians, blacks and Asians all had different "disabilities," and it is not possible to distill from Walker's writings a straightforward "ranking" of the three. He was most inclined to use the concept of "survival of the fittest" with respect to Indians, who in his view faced the choice of assimilating into white American culture or disappearing.[85] Although Walker saw blacks and Asians as inferior, he seldom discussed them explicitly in terms of a competition between races. He seems to have subscribed to what Livingstone calls "social Lamarckism," whereby in earlier phases of human history, isolated groups of people evolved, in response to different natural environments, into the different racial "stocks." These stocks could then evolve (or devolve) further in response to contact and competition with other stocks and as a result of education and other human institutions, but their long inherited basic characteristics would make it impossible for them to change too much. Most importantly, Walker implicitly held, with many other northeastern elites, that racial interbreeding was deleterious to all concerned, as it "diluted" the various stocks and thus made it impossible for each stock to develop it unique gifts to the fullest.[86] The racial stocks at the top of the hierarchy were the Americans, who had descended from those British of pure "Anglo-Saxon" blood, itself the best sort of "Teutonic" blood (see below).

Walker's understanding of basic racial categories assigned to each race a specific level of mobility – not surprisingly, the less advanced races were less mobile, and less productive. The issue of racial mobility continued to be important even after his political economic thinking shifted to an emphasis on productivity. This was because his vision of the social body involved not only assumptions about what explained more or less acceptable *individual* behavior, but also definite ideas about which *groups* belonged where within the national territory. He believed that blacks were unlikely to spread throughout the American national territory (a possibility that troubled many northerners, even abolitionists, during and after the Reconstruction era). Walker argued in 1891 that the freed slaves "represented a race bred under tropical conditions, and could move up the mountainside or go northward only at a large sacrifice of vitality and force." Although the desire to escape from bondage, and sometimes the movements of their white masters, had caused them to spread over large parts of the territory, "we should expect to find [in the results of the Eleventh Census] that, during the twenty-five or twenty-seven years since the blacks were left free to move within the country upon their own impulses,

[85] F. Walker, "The Indian question," *North American Review* 66, 239 (1873), 329–388.
[86] *Ibid.*, 385; Livingstone, *Nathaniel Southgate Shaler*, 79–122.

social, economical, and climatic forces have been operating to redress the balance."[87] Sixteen years earlier, he had made a similar argument in response to the agitation by Westerners against the "yellow peril" of Chinese immigration: "Of the natives of the Celestial Empire who cook and wash for our people, very few have yet ventured across the Rocky Mountains." Eastern white Americans should rest easy despite Western scaremongering, since "masses of foreign population thus unnaturally introduced into the body politic, must sooner or later disappear, like the icebergs that drift upon the currents of our temperate seas . . . their only part, endurance; their only end, extinction."[88]

Taken as an aggregate, "foreigners," too, were relatively immobile. The results of the 1870 and 1880 Censuses played a major role in convincing Walker of this "fact." In text accompanying charts which he inserted in the published results of the 1880 Census, and in his contribution to the ninth edition of the *Encyclopaedia Britannica* on "political geography and statistics," Walker made much of the confinement of both blacks and foreigners within certain well-defined geographical bands of latitude, altitude and even rainfall and average temperature.

The foreign-born have settled mainly between the 38th and 45th degrees of latitude . . . The colored people, all but an inconsiderable fraction, live between the 29th and 40th degrees of latitude. In this respect, the foreign and colored elements are largely complementary . . . Only 14.10 per cent. of the foreign population live in the regions raised 100 to 500 feet above the sea, while within those regions are found not less than 44.95 per cent. of the colored population.[89]

The charts in the 1880 Census results forming the basis for these observations are also accompanied by detailed breakdowns of the significant "racial" elements in terms of the recognizable physiographic regions of the national territory (see Figure 10 for an example).[90] Although Walker undoubtedly considered "foreigners" a separate racial category *before* absorbing the results of the two censuses he oversaw, there is no doubt that the regional grid of specification provided by the censuses made it far easier for him to conclude that his preconceptions were vindicated, to reify foreigners into a seemingly scientific biological category (see Chapter 5).

In contrast to these more or less immobile races, Walker ascribed to native whites a comprehensive mobility. The native white man "represented a race

---

[87] F. Walker, "The colored race in the United States," *The Forum* 11 (1891), 501–509, reprinted in *DES*, vol. II, 127, 130.

[88] F. Walker, "Our domestic service," *Scribner's Monthly* 11 (1875), 273–278, reprinted in *DES*, vol. II, 236–237.

[89] F. Walker, "Part III: Political geography and statistics," in "United States," *Encyclopaedia Britannica*, 9th ed. (New York, 1888), vol. XXIII, 820–821.

[90] Office of the Census, *Statistics of the population of the United States at the Tenth Census*, F. Walker and C. Seaton (comps.), (Washington, DC, 1883), XLII–LXXVI.

bred in the northern latitudes, and was hence thoroughly at home on the mountain side or table-land; while yet, by the privilege of his strain he could, without danger or great inconvenience, move southward if his interests required."[91] This was a convenient state of affairs from the perspective of manifest destiny, as it naturalized the spread of "native Americans" over the whole continent.

Walker's views on the "rigidity" or permanence of racial categories were neo-Lamarckian. For him environment (which represented fluidity and fuzziness of racial traits) and blood (which represented fixity) interacted in a variable fashion. The positive influences offered by the American environment (particularly the climate) proved perfectly suited for those of "Teutonic" stock, and in fact improved it. Other races were more or less impervious to American influences, or susceptible only to the negative environmental influences associated with crowded industrial cities. This helps to explain the significance Walker placed upon foreign *parentage* as well as foreign birth. If environmental influences could act quickly, unmuted by racial inertia, the provenance of one's parents would be less relevant to one's prospects.

For some races, particularly the "beaten races" from Eastern and Southern Europe, burdened with a long legacy of despotic rule and serfdom, the relevant environments had had their say long ago and in a negative direction, and had produced a stock ill equipped to reap the benefits of the New World.[92] For other races such as the English, the "most ingenious branch" of the "great inventive Teutonic race," the American environment encouraged further advancement, and had in fact already led to the formation of a new "American" race.[93] The early settlers

constituted ... a picked population. The possibilities of gain which reside in breeding from the higher, stronger, more alert and aggressive individuals of a species are well recognized in the case of domestic animals; but there have been few opportunities of obtaining a measure of the effect that could be produced upon the human race by excluding from propagation the weak, the vicious, the cowardly, the effeminate, persons of dwarfed stature, of tainted blood, or of imperfect organization.[94]

The demanding context in which labor-saving was of paramount importance had forged a distinctively American inventiveness among these already advantaged early settlers, "and through such incessant practice, originality of conception, boldness in framing expedients, and fertility of resource grew by exercise in father and mother, and were transmitted with increasing force to sons and daughters, till invention came to be 'a normal function of the American brain'."[95] In his famous 1891 essay on "immigration and degradation,"

[91] Walker, "The colored race in the United States," 128.

[92] F. Walker, "Restriction of immigration," *Atlantic Monthly* 77 (1896), 822–829, reprinted in *Discussions in economics and statistics*, vol. II, 438.

[93] F. Walker, "American manufactures," *Princeton Review* 11 (1883), 213–223, reprinted in *DES*, vol. II, 182.   [94] *Ibid.*   [95] *Ibid.*

Walker explained that the climate of the United States had allowed the nation "to take the English man and improve him . . . adding agility to his strength, making his eye keener and his hand steadier, so that in rowing, in riding, in shooting, and in boxing, the American of pure English stock is today the better animal."[96]

Walker believed the benefits of the New World could extend to other European "races" as well, particularly if they could make their way to the agricultural hinterlands upon arrival. In 1874, he wrote that "many a wretched beer-guzzler hanging about the saloons of New York, Philadelphia, and Boston . . . would, had he once been carried through Castle Garden and dropped five hundred miles beyond New York, have become a useful and prosperous citizen."[97] Similarly, "the Germans, the Scandinavians, and, though in a lesser degree, the Irish and French Canadians, who have made their homes where they are surrounded by the native [white] agriculturalists, have become in a short time almost as good Yankees . . . as if they had been born upon the hills of Vermont."[98]

But again, this optimism extended only to people originating in Northern and Western Europe. As the major sources of immigraton shifted during the late nineteenth century from Northern and Western to Southern and Eastern Europe, Walker's assessment of its impact necessarily darkened in stride with the general opinion of northeastern cultural elites, and America's social problems came to seem ever more intractable.[99] He began to see the "tainting" of "blood" as the original evil from which almost all others flowed. Pauperism, for example, was intrinsic to paupers:

Those who are paupers have the pauper taint; they bear the pauper brand . . . [whereas] almost nothing can push the poor who are not of the pauper type across the line of self-support, and keep them there, so long as the spirit of independence exists in the community [read "race"] to which they belong. Beaten down by misfortune, no matter how sudden and terrible, they reassert their manhood and reappear on the side of those who owe, and will owe, no man anything.[100]

Here, the self-made man turns out to be "made" by his blood as well as by his own efforts.

Walker's racial thinking brought together many currents then popular among the educated elite: a touch of social evolutionism, a well-developed rhetoric of "Teutonic origins," now and then a reference to the phrenology associated with "scientific racism," and a neo-Lamarckianism which gave

---

[96] F. Walker, "Immigration and degradation," *The Forum* 11 (1891), 634–643, reprinted in *DES*, *v.2*, 426.
[97] Walker, "Occupations and mortality of our foreign population, 1870," 216–217.
[98] F. Walker, "American agriculture," *Princeton Review* 9 (1882), 249–264, reprinted in *DES*, vol. II, 162–163.   [99] Solomon, *Ancestors and immigrants*, 44.
[100] F. Walker, "The causes of poverty," *The Century* 55 (1897), 210–216, reprinted in *DES*, vol. II, 456.

different kinds of "blood" variable responsiveness to the American environment.[101] David Livingstone's verdict on Walker's contemporary Nathaniel
Shaler (professor of geology at Harvard and, like Walker, a prominent public
voice on many issues) would suit Walker as well: he "made few original contributions to the study of race."[102] But Walker's views had some influence.
Walker did contribute significantly to the rising alarm about immigration, and
his "scientific" authority as two-time census Superintendent lent the fledgling
anti-immigration movement considerable added legitimacy. In particular, his
seemingly insignificant addition of questions about "foreign parentage" to the
1870 and 1880 Censuses ended up giving numerical form to a colloquial distinction between "old stock" and "new stock" immigrants, and would thereby
prove a powerful anchor for immigration restriction laws in the early twentieth century (see Chapter 7).

To sum up, Walker was not fundamentally concerned about Indians, blacks
or Asians, trusting their limited numbers or limited environmental range to
keep the northeastern United States free from invasion by these races. His
major worry was immigration, particularly from Eastern and Southern
Europe. Through his construction of census results, Walker had come to view
"foreigners" as a biological racial category, and although he recognized
regional and national subdivisions of foreignness, the subdivisions retained an
assumed biological character. Eastern and Southern European immigrants
had been beaten down to the point of unsalvageability by generations lived
under despotic rulers and backward economic systems. They were at the root
of many of the problems threatening to destabilize America's productive
system (class struggle and all the attendant strikes, riots and socialist agitation, as well as suboptimal productivity and mobility) and thus also the chief
threats to the American exceptionalism Walker so treasured. They made the
American working population less manly than it should have been, preventing the development of a brotherhood of men that would have rendered class
divisions superfluous. These immigrants were immobilized by their disadvantages, but not safely in the agricultural South like blacks, or far to the West
with Asians and Indians. They were immobilized right in the midst of the
urban-industrial core of the booming American economy, right in the midst
of the lives and jobs of the native white American men on whom all possibility of national glory depended. Walker found this alarming in the highest possible degree, not only because of the disruptive effects these foreigners had on
production, but perhaps even more because of the threat they posed to
national *reproduction*. Their "tainted blood" not only led them to act in ways
injurious to native white Americans, but also threatened the long-term survival of native white blood.

---

[101] See T. Gossett, *Race: The history of an idea in America* (New York, 1963), chs. 4 and 5, for a
useful overview of racial thinking during Walker's era.
[102] Livingstone, *Nathaniel Southgate Shaler*, 157.

Immigrants and population growth rates

Since before Adam Smith, the issue of population growth had been an important one in European writings on political economy. It was important not merely in the Malthusian sense of its relationship to the rate of accumulation of wealth, but also as an indicator of national vigor or health. It is not surprising that in the "mercantilist" phase of capitalist development, dominated as it was by the question of comparative *national* wealth, the sizes and growth rates of the national populations which produced that wealth should have interested governments and educated citizens greatly.[103] In this sense, political economy had by Walker's time long involved questions of *reproduction* as well as of *production*. It is not an exaggeration to say that for Walker, population growth rates constituted the single most important method of comparing the well-being of different nations, and thus the chief means by which the fate of his American exceptionalism could be discerned. His strong national chauvinism, coupled with his belief in the authority of statistical demonstration, gave him a heavy psychological investment in the fortunes of population growth rates in the United States. And, I would argue, this chauvinism is at the root of the unusually virulent anti-immigration stance he eventually took.

In the previous section of this chapter, I noted that even during the mid-1870s, when Walker still saw economic benefits in immigration, he could be provoked to condemn it by the suggestion that the American economy depended on the influx of Europeans. My hypothesis is that Walker's patriotism was anchored in a projection of his manhood ideal onto the country as a whole: the last thing a great nation could be, just as the last thing a full-blooded man could be, was dependent on other nations or men. If population growth rates were an important measure of national greatness, and the United States showed remarkable population growth rates, then the possibility that those growth rates were dependent on immigration rather than indigenous growth was a grave threat to American national pride. This threat, I believe, was one of the motivations that led Walker to make so much of the distinction between "native whites" and "foreigners," and to biologize the latter category in his construction and interpretation of census results. The specter of demographic dependence haunting Walker's national pride required him to demarcate native whites as the only true Americans, and to invest heavily in the fortunes of the growth rate of this group relative to that of "foreigners." Although his main concern was with labor, he certainly saw even his and Mrs. Walker's five children as relevant to the struggle. In 1878, he proudly wrote to his friend Daniel Coit Gilman that "Mrs. Walker has just made another contribution to the Census of 1880 – a boy, native, white, and, I am

[103] McNulty, *The origins and development of labor economics*, 27–28.

free to say, a great credit to his parents."[104] Walker, after all, had an "extraordinarily homogeneous" ancestry, almost all of his forebears "came over in the first great wave of English immigration before 1650, and there was afterwards little or no admixture from other than British stock."[105] He undoubtedly felt the need to vindicate native white Americans at a personal as well as at a demographic level. He was a true champion of "national manhood."[106]

In this final section of the chapter, I will trace the changes that occurred in Walker's attempts to explain a disturbing fact revealed by the succession of censuses: that since at least the Civil War, the growth rate for native whites had been falling off, while the growth rate of immigrants had been consistently higher. In articles appearing occasionally, from the early 1870s through to the early 1890s, Walker returned again and again to this sequence of census results, and each time tinkered with his explanation, until in 1891 he had arrived at an account that satisfied him. Each time he took up the subject, he began by reviewing the remarkable predictions made by one Elkanah Watson in 1815. Watson had simply computed the typical decadal growth rate revealed by comparison of the 1790, 1800 and 1810 Censuses and extrapolated onward through each decade to 1900. These predictions were remarkable not for their method but because they had proven so accurate up to 1860, deviating from the actual figures by a trifling percentage each decade.[107] With the 1870 results, however, the rate for native whites slows, and overall population growth is only buoyed by the addition of rapidly growing numbers of immigrants. Dependency on Europe was certainly one way to interpret these numbers, but Walker was grimly determined to find another explanation.

In 1873, Walker was not yet much alarmed by the fortunes of Watson's prognostications, but his disapproval of the trend was already clear. Up until the middle of the nineteenth century, the American population had been in the best possible situation to allow high growth rates: a rapidly expanding agricultural frontier had posed no obstructions to high fertility (it was to confirm this association that Walker included a map of fertility in the 1874 *Statistical Atlas* – see Chapter 5). The change is attributable not so much to the Civil War, which, all told, could have accounted for losses of roughly 2.5 million. Instead, "it began when the people of the United States began to leave agricultural for manufacturing pursuits; to turn from the country to the town; to live in up-and-down houses, and to follow closely the fashion of foreign life." And it will continue,

[104] F. Walker to D. Gilman, Dec. 4, 1878, Daniel Coit Gilman Papers, Special collections, Johns Hopkins University Library, MS 001.

[105] J. Munroe, *A life of Francis Amasa Walker* (New York, 1923), 5.

[106] D. Nelson, *National manhood: Capitalist citizenship and the imagined fraternity of white men* (Durham, NC, 1998).

[107] F. Walker, "Our population in 1900," *Atlantic Monthly* 32 (1873), 487–495, reprinted in *DES*, vol. II, 30–32.

as the line of agricultural occupation draws closer to the great barren plains; as the older Western States change more and more to manufactures and to commerce; as the manufacturing and commercial communities of the East become compacted; as the whole population tends increasingly to fashion and social observance; as diet, dress, and equipage become more and more artificial; and as the detestable American vice of "boarding," making children truly "encumbrances," and uprooting the ancient and honored traditions of the family, extends from city to city and from village to village.[108]

Immigration figures in the main text of this article are presented by Walker not so much as the immediate source of these developments but as a demographic screen obscuring their effects by artificial compensation.[109] Yet in an endnote, Walker foreshadows his later arguments:

The popular notion that the relative decline in the national increase has been due to a loss of physical vigor, will not bear the test of evidence. At the time when our population was purest, when immigration was so slight as to be hardly appreciable, the American people had shown the capability of maintaining a rate of increase which should double their numbers in twenty two years; and this, over vast regions and through long periods.[110]

When Walker again takes up Watson's numbers in a contribution to the ninth edition of the *Encyclopaedia Britannica* written during the mid-1880s, he gives foreigners a more prominent role, singling out "the introduction of a vast body of peasant blood from Europe" as an important factor, though he does not specify the exact mechanism by which it lowers native white growth rates. The increasing presence of women in the waged workforce is another relevant trend.[111] In 1889, in an article entitled "Growth of the nation," Walker reiterated the factors mentioned in earlier writings, but elevated the endnote quoted above about loss of vigor to a place in the main body of his text (verbatim). On the other hand, there is no mention of "peasant blood," the focus instead being on the rise of manufacturing and a new penchant for luxurious living. As a result of these changes in particular, "[t]he objects which, during the earlier period, the nation had pursued with such singleness and eagerness of purpose [presumably reproduction and farming], were thereafter to divide their thoughts and energies with other objects proper to a fuller stage of development."[112]

It is in two papers published in 1891 that immigration eclipses other factors in Walker's explanations. In "The great count of 1890," he works through the usual culprits (manufacturing, urbanization, "boarding," luxurious dissipation), but his treatment of immigrants takes on a new gravity:

---

[108] *Ibid.*, 42–43.   [109] *Ibid.*   [110] *Ibid.*, 44, note 1.
[111] Walker, "Political geography and statistics," 818.
[112] F. Walker, "Growth of the nation," from Phi Beta Kappa oration, Brown University, June 18, 1889, reprinted in *DES*, vol. II, 204–205.

Finally, vast hordes of foreigners began to arrive upon our shores, drawn from the degraded peasantries of Europe, accustomed to a far lower standard of living, with habits strange and repulsive to our people. This, again, caused the native population to shrink within themselves, creating an increasing reluctance to bring forth sons and daughters to compete in the market for labor.[113]

Walker's most influential article on population growth rates, entitled "Immigration and degradation," also came out in 1891. In it he drew together all the threads of his previous arguments, and finally specified exactly how foreigners slowed native white population growth. Tellingly, he begins by excoriating the "grotesquely wrong" claim that American population growth is indebted to European immigration, an assertion "in the highest degree derogatory to the vitality of our native American stock, and to the sanitary influences of our climate."[114] But the temporal coincidence between burgeoning immigration and diminishing rates of population growth for native whites is too close to dismiss. As in the other 1891 article, Walker claims that the explanation lies in a process whereby "the native population more and more withheld their own increase." The flood of foreigners "constituted a shock to the principle of population among the native element. That principle is always acutely sensitive, alike to sentimental and to economic conditions." By "the principle of population," Walker meant the willingness of men to engage in sexual intercourse; the presence of foreigners was a threat to white American male potency. That Walker had men in mind is implied in his image of precisely how the "shock" was produced:

Our people had to look upon houses that were mere shells for human habitations, the gate unhung, the shutters flapping or falling, green pools in the yard, babes and young children rolling about half-naked or worse, neglected dirty, unkempt . . . But there was, besides, an economic reason for a check to the native increase. The American shrank from the industrial competition thus thrust upon him. He was unwilling himself to engage in the lowest kind of day-labor with these new elements of the population; he was even more unwilling to bring sons and daughters into the world to enter into that competition.[115]

It would mostly have been men who would have been in a position to witness such debilitating scenes, and it was after all men who in Walker's mind properly decided when and how often to have sex. The sexual imagery of shrinking and withdrawing in these articles is none too subtle, and neither is it too much of a stretch to say that Walker feared the onset of a sort of national impotence among native white working men. His great consolation was that he felt he had definitively proven the negative effects of immigration. Indeed, he went on toward the end of the article to claim that had the foreigners never

---

[113] F. Walker, "The great count of 1890," *The Forum* 11 (1891), 406–418, reprinted in *DES*, vol. II, 122.   [114] Walker, "Immigration and degradation," 417.   [115] *Ibid.*, 424.

come, native whites would have reproduced at a much higher rate, high enough to more than make up for the absence of immigrants.[116] Thus, far from being dependent on Europe, the United States had actually suffered from immigration. If American exceptionalism were after all to prove unfounded, it was a fate that could not be laid at the feet of true Americans, but rather of immigrants. With this, Walker rested his case, though not his activism, as we shall see in Chapter 7.

In this chapter, I have attempted to spell out Francis Walker's explicit understanding of American society, the ideology, if you will, through which he would interpret the results of the census. The second moment of the cycle of social control, which I have followed Foucault in calling "normalizing judgment," is that which involves bringing knowledge about the body to be governed into contact with norms of health or well-being. The conclusions reached at this stage are then filtered through the governmental injunction not to over-regulate, in order to decide finally "what is to be done" (see Chapter 7). Walker understood the statistical knowledge he had been so instrumental in producing through a well-developed political economic theory. Lent authority by statistical "facts" he himself had constructed, and infused with a strong (if in some respects vague) racism and an even stronger national chauvinism, Walker's political economy allowed him to make a case that the overriding cause of America's departures from norms of national health was immigration from Eastern and Southern Europe. In the sphere of production, his shift to a focus on productivity trained attention on the inherent characteristics of laborers, which could in turn easily be attributed to race or national origins; in the sphere of reproduction, his conviction that the temporal coincidence of falling native white growth rates and rising immigration signaled a causal relation had a similar effect. All of this was made possible by a painstaking naturalization of the category of "foreignness" in a nation composed almost entirely of descendents of immigrants (in retrospect, this was no mean feat), and by the reification of "national" growth rates as meaningful statistical aggregates.

The hope and promise of America's economic success was the vigorous white male worker, making use of his mobility and natural productivity to engage employers in competition for wealth beneficial to everyone, and thereby to help forge a trans-class, national community of productive men. Yet with rising immigration, this worker was surrounded ever more often by immobile, stagnant pools of inferior labor, slowing competitive growth with bad work habits, hawking socialism, anarchism and other injurious doctrines, and keeping wages artificially low to the detriment of all. Partly because of artificially low wages, women were leaving their proper place at home to find wage work more often, and even agitating for political independence. Worse,

---

[116] *Ibid.*, 425.

the mobility of the intrepid native white worker brought him into contact with scenes of squalor and degradation that dampened his enthusiasm for propagating his precious seed. In 1790, before this hideous onslaught, "both the social traditions and the religious beliefs of the people made the will of the husband supreme. While this law reigned in the household, there was nothing to repress, but, on the contrary, everything to encourage, a rapid growth of population."[117] Men could be men on their own terms. Now, obstacles were legion, both to efficient production and to vigorous reproduction.

[117] Walker, "Growth of the nation," 197.

7

Manhood, space and governmental regulation

Setting the stage for Francis Walker's program in Part One, and following him through the "observational" and "judgmental" moments of its cycle of social control, have taken me quite far from the theoretical discussion of governmentality in Chapter 1. Along the way, the work of Foucault has been most useful, first, in providing the framework for recasting Walker's subjectivity as a symptomatic tangle of threads in the weave of a larger discursive formation (Part I), and second, in encouraging a focus on the often spatial politics of social construction whereby the American social body was constituted. In short, Foucault has made it possible to portray both subject and object of governmentality as constructions, and to accent the links between construction and control. But there is nothing definitively "governmental" about this. The most distinctive moment of any governmental program, that which sets it off as different from the colonial or invasive bureaucratic styles of administration that have been the focus of most critical theorizing about state power, is the moment of regulation. It is here that Foucault's thoughts on governmentality are most useful in a straightforwardly theoretical way: Francis Walker's program follows the logic spelled out in Chapter 1 with almost uncanny precision.

In the present chapter, as I connect Walker's "normalizing judgments" on the American social body with his ideas on what was to be done, it will finally be possible to say, "*these* features of Walker's thought can only be adequately grasped through the concept of governmentality." Since this book is fundamentally *about* governmentality (more than it is about Francis Walker or late nineteenth century American state formation), the present chapter is of central importance. Its purposes are three: to illustrate what the logic of governmental regulation might entail at a national scale; to show how the concrete ideological environment of this logic, outlined in Chapters 4 and 6, inevitably inflected it; and finally to highlight the spatial features of national governmental regulation. After laying out Walker's general attitude toward national regulation, I will discuss the first major pillar of his regulatory program: the struggle to redesign education for modern manhood. Next, I will briefly explore the spatial logic that compelled him to supplement the educational program with some sort of spatially restrictive regulatory policy, and present a brief summary of his policy toward American Indians in order to make clearer by contrast exactly how his advocacy of immigration restriction addressed the dictates of this spatial logic. Immigration restriction represents the culmination of his governmental program, both in the sense that it brings to fruition all of the spatial aspects of his thinking, and because it was his most important legacy to the twentieth century. Thus, I will round out the chapter, and the narrative as a whole, with an account of Walker's influence on immigration policies in the early twentieth century.

Walker's general approach to governmental regulation

As noted in Chapter 1, a key stage in the genealogy of governmental thinking was reached when early political economists began to see national economies and societies as entities distinct from their rulers and animated by their own laws. In so far as sovereigns accepted this view, the goal of government policies began to be that of learning about the laws of society and economy in order to decide how much and how best to regulate. It would be easy here to misunderstand my claims for Walker's distinctiveness. I do not mean to suggest that Walker was the first or only person to *call for* empirically based regulation. However, he *was* one of the first prominent officials to move beyond "gestural governmentality" to the actual use of statistics as a basis for decisions about whether and how to govern. The literature on administrative capacity simply does not capture these nuances.

Public figures like Walker, imbued with what we can retrospectively recognize as a governmental mentality, also had a more sensitive conception of regulation itself. Displaying the subtlety later highlighted by Foucault and others, they were able to distinguish between *restrictive or coercive* regulation and *supportive or fostering* regulation. This in turn indicates a more subtle *politics* of regulation. Just as the moment of observation has a specific politics centered on visibility and orderliness, and the moment of judgment a more obvious politics (in Walker's case, of race, class and gender), the moment of regulation has its own implicit but contestable distribution of authority and submission. Briefly, in an ideal governmental view, there is a reciprocal relation of authority between the governing and the governed. The social body is made to reveal its laws without explicit "consent" at the moment of observation (see Chapter 5), but then the government is obliged to respect the integrity of those laws and refrain from excessive intervention at the moment of regulation. Finally, however, if the government follows this stricture, the society is in turn obliged to live by the regulations deemed necessary by the authorities. This political sequence is only implicit in Foucault's writings on governmentality, but it is crucial for any understanding of why governmental programs so often fail in concrete settings. More so than in the more familiar parallel process of "democratic" politics, the politics of governmental regulation (especially at a national scale) requires trust on the part of the citizenry. In so far as legislators make laws with reference to the interests of the constituencies by which they are elected, there is no great mystery in the process from the latter's perspective. But in so far as laws or administrative regulations come to be based on social scientific considerations of the wider national social body, citizens may not see the point, and thus must have some degree of faith in the government. For this reason, the political bargain specific to governmentality usually remains an idealized abstraction, and concrete programs of governmental regulation must deal regularly with all manner of challenges from the governed.

In general, Walker's views on the *principles* of social regulation did not change significantly over the course of his public career, though his 1890s advocacy of immigration restriction could be read as a move toward more aggressive control *in practice*. His steadiness in this respect should not be surprising in light of the fact that his first high-profile public position was as superintendent of the 1870 Census. This experience probably established the empiricism that was so central to his understanding of social policy, and predisposed him in all of his subsequent roles to proceed on the basis of concrete information as often as possible. Yet even in his late, bare-knuckled polemic against open immigration, Walker displayed what might seem, lacking an understanding of governmentality, to be a surprising timidity in recommending actual regulation. His writings are peppered throughout with expressions of an aggressive urge to put things in order, to clean up, purify and set straight the American social body. But, attentive to the need for care and circumspection, he obeyed as a governmental subject the same self-control exercised more generally by the "ideal man" of the late nineteenth century (see Chapter 4). The prevailing model of manhood, in which a portion of one's vital force should be expended on reining in its more untrammeled and aggressive expression, might well have been the model through which Walker understood his own approach to regulation. The tenor of his writings often suggests that, on certain issues, his faithfulness to an ethic of circumspect regulation cost him considerable expenditures of willpower.

Toward the end of his most sophisticated essay on regulation, entitled "Socialism" (1887), Walker captures perfectly the governmental attitude:

May we not believe that there is a leadership, by the state, in certain activities, which does not paralyze private effort; which does not tend to go from less [regulation] to more; but which, in the large, the long, result, stimulates individual action, brings out energies which would otherwise remain dormant, sets a higher standard of performance, and introduces new and stronger motives to social and industrial progress?[1]

In much the same way that his political economy revolved around an insistence on social realities, his regulatory program revolved around a commitment to confront minimalist regulatory ideals with the facts of widespread and unavoidable regulation, and to invest these facts with some of the trappings of principle. The very next passage in "Socialism" locates the burden of proof in debates over the proper extent of regulation:

[W]hile I repudiate the assumption of the economic harmonies which underlies the doctrine of *laissez-faire*, and while I look with confidence to the state to perform certain important functions in economics, I believe that every proposition for enlarging the powers and increasing the duties of the state should be long and closely scrutinized; that a heavy burden of proof should be thrown upon the advocates of every such scheme; and that for no slight, or transient, or doubtful object should the field of industrial activity be trenched upon in its remotest corner.[2]

Leaving aside for the moment exactly what Walker considered minimally necessary state intervention, it must be stressed that his distrust of schemes for regulation did not amount to automatic disapproval. Characteristically, he insisted that since "in every civilized country the functions of government have been pushed beyond the mere police powers," the question of how much to regulate was a complicated, empirical question.[3] It was no use appealing to principles alone; it was necessary to begin with facts.

On the other side of the bargain, regulations made under the sign of proper circumspection were for Walker perfectly within the rights of states, and were to be obeyed without complaint in most cases. His desire to see citizens and their elected representatives *accept* this principle was at the heart of his politics of regulation, and is thus worth pausing to dwell upon. We have already seen Walker asserting the "right to learn whatever [the citizens, through the agency of the Census Office] might please to know" (see Chapter 5); he similarly considered the taxation necessary to maintain the state one of its basic "righteous demands."[4] Walker showed a consistent hostility toward citizen resistance against legitimate state demands, dismissing, for example, the appeal to "natural rights" as an argument against protectionism:

I do not deem myself qualified to say much about natural rights, having never lived in a state of nature, but having resided all my life in communities, more or less civilized, whose citizens were required to render numerous and onerous services, to refrain from

[1] F. Walker, "Socialism," *Scribner's Magazine* 1 (1887), 107–119, reprinted in F. Walker, *DES*, vol. II, D. R. Dewey (ed.) (New York, 1899), 270.   [2] *Ibid.*   [3] *Ibid.*, 251.
[4] F. Walker, "The bases of taxation," *Political Science Quarterly* 3 (1888), 1–16, reprinted in F. Walker, *DES*, vol. I, 86.

many courses agreeable to themselves, to make heavy contributions, to submit to severe sacrifices, to walk in paths instead of roaming at will over the fields – all for the general good.[5]

Walker was of course a nationalist, an inheritor of the federalist tradition that had produced the Constitution but then had succumbed by the early nineteenth century to the distrustful Republican localism associated with Jefferson.[6]

Walker linked the birth of American nationalism (or, as he termed it, "nationality") closely with the growth of federal state institutions and interventions. His writings on American political history nicely illustrate his abiding concern with the terms of the "bargain" outlined above. The last book he published was *The making of the nation, 1783–1817*, a subtle attempt to explain how a sense of national citizenship and nationhood emerged during the first thirty years of the Republic's existence.[7] Political history represents somewhat of a departure from the range of subjects with which Walker was normally associated. But in fact, it was a logical culmination of his publishing career, a reflective step back in which he placed his own struggles to get his country to accept governmental regulation in a longer historical frame. The narrative dwells on all the major events and circumstances which have become standards in early American political history (relations with France and England, the growing animosity between Hamilton and Jefferson as a harbinger of the emerging party system), but always with a special eye to the long and experimental process of expansion of actual federal powers. Thus, alongside such matters as the Constitutional Convention, the influence of George Washington, the Jay Treaty and the Alien and Sedition Act controversy, Walker recounts such things as the growth and westward spread of the white population from census to census, and the development of road and water transport networks. In a rudimentary way, he was arguing that what would today be termed social history is always relevant to political history: "we are now talking, not of names, but of things; not of written instruments of public declarations, but of real social and political forces."[8] Substantively, his chief claim was that social as well as political developments during the thirty years *after* the Constitutional Convention of 1787 were primarily responsible for forging a workable sense of nationhood. He thereby placed himself in explicit opposition both to the older tradition of political history which had focused narrowly on the Constitution as the source of national unity, and to the dominant late nineteenth-century school which saw American national unity as having remained a fiction until forcibly established by the Civil War.[9]

---

[5] F. Walker, "Protectionism and protectionists," *Quarterly Journal of Economics* 4 (1890), 245–275, reprinted in *DES*, vol. I, 114.

[6] See S. Elkins and E. McKittrick, *The age of federalism* (New York, 1993) for an exhaustive treatment of this crucial period.

[7] F. Walker, *The making of the nation, 1783–1817* (New York, 1897).   [8] *Ibid.*, 269.

[9] *Ibid.*, 268.

Walker had summarized the argument of the book in an 1895 essay for the *Forum*, and thus it is possible to pick out the points he considered most important. The Confederation period immediately after the Revolution had been characterized by a continued reluctance on the part of the new states to commit to political integration, with well-known results.[10] According to Walker, demographic expansion, and the growth of transport infrastructure formed an indispensable backdrop for the developments that changed this situation. One extremely important political step toward fulfilling the promise of the new Constitution written to strengthen the union was the appointment of Supreme Court justices who were strongly committed to national unity. Walker significantly singles out as especially important two constitutional principles established by John Marshall: (1) "that, while the General [Federal] Government is limited as to its objects, it is yet as to those objects supreme," and (2) that the federal government is free to use any means not expressly forbidden by the Constitution to attain those objects.[11] (Walker undoubtedly had in mind here his difficulties with census-taking.) The Louisiana Purchase was for him another crucial step in the birth of nationalism, "second only in our history to the adoption by the Constitutional Convention of Randolph's resolution," structuring the federal government in three supreme branches, because the Purchase required a significant expansion of state activity in general, and more particularly because it was settled by American citizens who under the Territorial system *grew accustomed to direct federal authority.*[12] Walker considered the War of 1812 to have "sealed" American nationalism, since it both galvanized the people in opposition to a foreign enemy and led, directly or indirectly, to a drastically expanded central state presence.[13] For Walker, the fact that these steps were taken under Republican administrations nominally opposed to big government was absolutely crucial:

No government can carry on a war without feeling strongly the impulse to aggrandize its own powers . . . It was a Republican administration which passed the first distinctly protective tariff, imposed excise duties, enacted a direct tax, rolled up enormous public debt, created a sinking fund, and founded a national bank. Who then was left to protest that the United States should not become a nation?[14]

The last line in this passage also closes his book, and is actually a profoundly important question for Walker, not merely a rhetorical flourish. By the second decade of the nineteenth century, the nation had consolidated itself not only in political terms but also in "racial" terms: the "old stock" would propagate itself relatively undisturbed by immigration until the Irish influx of the 1840s. It was

---

[10]   *Ibid.*, 2–20.
[11]   *Ibid.*, 167, 253; F. Walker, "The growth of American nationality," *The Forum* (1895), in Francis A. Walker Papers, 1862–1897, Massachusetts Institute of Technology Archives and Special Collections, MC 298, Box 3, 395.   [12]   Walker, *The making of the nation*, 184.
[13]   *Ibid.*, 254–256; Walker, "The growth of American nationality," 397–399.
[14]   Walker, "The growth of American nationality," 400.

this period that Walker used as a baseline in interpreting Elkanah Watson's projections of American population growth, and in constructing his own dire analysis of the effects of later immigration (see Chapter 6). Walker was in effect implying that to explain subsequent challenges to a general nationalist consensus, one had to find culprits other than the descendants of the "old stock" citizens. In so far as class struggle and more diffuse resistance to federal authority appeared in later decades, they had to be attributed to "foreign elements."

Walker's closing passage is also pregnant with significance as an historical comparison. The comments about the effects of war, as he was doubtless well aware, could also have described the national state's situation in the post-Civil War era when Walker was cutting his administrative teeth.[15] The struggle to get the nation to accept its expanded federal state was for Walker both a general and a specifically governmental issue. He made the latter clear in his opening address as Président-Adjoint to the International Statistical Institute in Chicago in 1893, where he pointedly and almost enviously welcomed delegates from countries "in which both legislation and administration freely acknowledge their obligation to consider and to defer to the results of statistical inquiry."[16] Not surprisingly, Walker saw the effort to educate citizens in this duty of obedience as itself a legitimate form of state activity. True to the general pattern characteristic of modern power/knowledge at whatever scale, one goal of regulation was, for Walker, improved regulability.

In broad outline, then, Walker quite fully inhabited what we can now identify as a governmental mindset with respect to regulation: he understood the need for caution, he saw the distinction between coercive and "fostering" regulation, and he was sensitive to the politics implicit in any national governmental program. The particular kinds of regulation appropriate for the American social body are most carefully discussed in the "Socialism" essay. I will give a brief overview of his arguments on specific regulations before turning to his educational program.

One of the most interesting features of Walker's writings on regulation is the care he took in distinguishing between "isms" and "ists." He seems to have hit upon this formulation in the late 1880s as a strategy for removing some of the acrimony from public debates; he recognized a pattern that remains patently obvious today, namely that acrimony tends to produce polarization and hence simplification of the issues under discussion. Since his governmental program was at one level a challenge to the simplifying effects of argument from principle alone, it is not surprising to find him groping for some way to get beyond the prevalent pattern. In discussing both socialism and protection-

---

[15] See R. Bensel, *Yankee leviathan: The origins of central state authority in America, 1859–1877* (New York, 1990), and G. Fredrickson, *The inner Civil War: Northern intellectuals and the crisis of the Union* (New York, 1965).

[16] F. Walker, "Address to the International Statistical Institute," from *Bulletin de L'Institute International de Statistique* 8 (1895), xxxvi–xxxix, reprinted in *DES*, vol. II, 141.

ism, Walker stressed repeatedly that one could advocate particular "socialist" (or "protectionist") measures without necessarily therefore *being* a socialist (or protectionist). He further separated the issue of whether a measure is socialistic or protectionist from the question of whether it has merit (in part, again, by noting that most "civilized nations" employ at least some such measures). If one begins with a prejudice in favor of minimal police powers and "regards all acts and measures enlarging the functions of government beyond this line as more or less socialistic . . . [one can no longer] use that term as one of contumely or contempt."[17]

The particular inspiration for this set of distinctions was in all probability Walker's former colleague at Yale, William Graham Sumner. Sumner, America's foremost champion of social Darwinism during the 1880s, was so strident and dogmatic in his condemnations of "socialism" and "protectionism" as to label them *morally* wrong, which of course implicated the character of the individuals who espoused such policies.[18] Walker undoubtedly perceived that Sumner's stridency and unwillingness to make fine distinctions encouraged his opponents to follow suit. The larger significance of Walker's clever response is twofold. First, by focusing the question of socialism on individual measures and detaching it from the question of merit, it created space for regulatory decisions to be made on an *empirical* basis. Second, in distinguishing advocates of particular socialistic measures from "socialists" properly so-called, this strategy went some way toward abstracting regulatory questions from matters of personal investment. As I indicated in Chapter 3, public men of affairs during the late nineteenth century were still partly immersed in a culture which offered no established way of depersonalizing one's intellectual positions. For early social scientists, there were as yet no clearly established and universally acknowledged hypothetico-deductive procedures for guarding against bias in scientific reasoning, and little of the accompanying culture of pride in self-erasure that subsequently became an important part of academic research. Thus Walker's new rhetoric can be seen as an interesting transitional moment in the larger shift from the era of public men to that of experts.

To return to the main thread of the argument, the basic police powers, in Walker's eyes, included "all that is necessary to keep people from picking each other's pockets and cutting each other's throats, including, alike, positive and preventive measures." Also included are "adjudication and collection of debts" and "the punishment of slander and libel," as well as "the war power."[19] A number of other measures are necessary, even if in some cases socialistic: protection of religious freedom, popular education (socialistic), construction and maintenance of bridges and roads (also socialistic), "repression of obtrusive

---

[17] Walker, "Socialism," 251.
[18] R. Hoftsadter, *Social Darwinism in American thought* (Boston, 1992 [1944]), 54, 63.
[19] Walker, "Socialism," 253.

immorality" and urban sanitary regulation (also socialistic).[20] Walker's explicit criteria for determining that a measure is socialistic were (1) that it could not be construed as serving basic public order, and (2) that, once established, it would tend further to enlarge state intervention, and beget dependency among the citizenry. Unfortunately, he had less to say about *precisely which* socialistic measures beyond the necessary few listed above were nevertheless acceptable. Later in the same essay, he catalogues a number of important socialistic measures (trade protectionism, labor legislation, state takeover of transport and communication infrastructure and regulation of industrial corporations) without offering verdicts on their advisability.[21] He clearly saw some point in protecting infant industries, preventing child labor and overseeing transportation, but in each case he also saw real dangers: the "pauperization" or permanent dependency of protected industries, the escalation demands by labor for such "unreasonable" measures as the eight-hour day, and (undoubtedly) the danger of corruption so clearly demonstrated in the Credit Mobilier scandal of the 1870s.[22] He was hesitant to regulate corporations, but contemptuously dismissed the argument that such regulation would unfairly restrain competition, noting that "the very institution of the industrial corporation is for the purpose of avoiding that primary condition [the vulnerability of capitalists to loss of their invested resources] upon which, alone, true and effective competition can exist."[23]

In his thinking about regulation, Walker was concerned above all else with promoting a self-sufficiency *modeled on ideal American manhood*. This meant, first, preparing men to cope with life as adults in industrial America, and second, protecting them from influences which could place them at a disadvantage from the very start. Although he had much to say about all the different sorts of regulation noted above, his strongest interests lay in those measures which he believed most effectively enhanced the manhood of white American men. In this manhood lay the key to the health of the American social body, and the hope for his American exceptionalism. The two social policy initiatives dearest to his heart were industrial and technical education, and immigration restriction. The former, though "socialistic" at the primary and secondary levels where it was undertaken by the state, was for Walker the most effective sort of "fostering" regulation; the latter, the most effective "coercive" or restraining regulation.

Education for modern manhood

On the strength of his public career, and the recommendations of such influential close friends as Daniel Coit Gilman, first president of Johns Hopkins

---

[20] *Ibid.*, 253–255.   [21] *Ibid.*, 260–265.
[22] Walker, "Protectionism and protectionists," 97–98, 106–107; F. Walker, "The eight-hour law agitation," a lecture delivered before the Boston Young Men's Christian Union, Mar. 1, 1890, reprinted in *DES*, vol. II, 379–396.   [23] Walker, "Socialism," 264–265.

University, Francis Walker was offered the presidency of the Massachusetts Institute of Technology while still in the midst of his labors as superintendent of the 1880 Census. He accepted, resigned his ongoing post as professor of political economy at Yale's Sheffield Scientific School in 1881, and ended up devoting much of his "vital force" during the last sixteen years of his life to building MIT. Inevitably, after a few years of experience, Walker felt qualified to begin writing on educational topics for the wider public, and his writings soon embraced educational policy in primary and secondary public schools as well as at the post-secondary level. He became a well-known, if unoriginal, advocate of industrial, technical and scientific education as a key means of preparing members of American society for the challenges of adult life in a new industrial age. The great inspiration for this movement in the anglophone world was Herbert Spencer, and two of its most famous proponents in America were Walker's friends Charles W. Eliot, president of Harvard University, and Daniel Coit Gilman of Johns Hopkins University.[24] Soon after taking the helm at MIT, Walker, too, was invited to give numerous addresses on such subjects. By 1885, his reputation was such that California railroad magnate Leland Stanford invited him to the west coast to advise him on founding a new university. Stanford was so impressed that he ended up offering Walker the first presidency of Stanford University on the spot (Walker politely declined).[25]

Walker's concern throughout, both in his writings and in his administration of MIT, was to foster a type of *manhood* appropriate for the times; as we shall see, his program for women was far more conservative. Walker actually had two models of manhood in mind: in his writings on public education at lower levels, the model citizen-worker; in his writings and activities as president of MIT, the model citizen-scientist. Although he was not explicit about it and probably did not see it in these terms, the two streams of his educational policy amounted to a vaguely Comtean vision of a technocratically ordered society supervised by a scientific elite. For Walker, the model citizen-worker was a creature who had been in abundant supply in the early stages of industrialization, when the chief source of labor had been farm families in the countryside. Growing up on the farm had endowed most men with the distinctive American qualities of energy and practical inventiveness that translated quickly and easily into "industrial mobility." (This assumption places Walker in the nineteenth-century lineage leading from Thomas Jefferson to Frederick Jackson Turner.) Once urban life became the norm for the majority of industrial workers, it became the task of a basic education to compensate for the absence of conditions for the natural cultivation of inventiveness and enterprise.[26] The

[24] A. Meyer, *An educational history of the American people* (New York, 1957), 238–241; F. Rudolph, *The American college and university: A history* (New York, 1968), 244–245.

[25] J. Munroe, *A life of Francis Amasa Walker* (New York, 1923), 307–309.

[26] F. Walker, "Manual education in urban communities," address before the National Educational Association, Chicago, IL, July 15, 1887, reprinted in F. Walker, *DE*, D. R. Dewey (ed.), (New York, 1898), 175–194.

elements Walker proposed to include in a basic industrial education at the secondary level were geometry, physics, mechanics, drawing and "shop work of one kind or another." Even young men who were not destined for the industrial labor force would gain from such a curriculum "the advantage derived from the training of the perceptive powers, the formation of the habit of observation, and the development of the executive faculty, the power, that is, of doing things as distinguished from thinking or talking or writing about them."[27] Walker was particularly concerned about this "executive faculty," lying as it did at the heart of his manhood ideal. As the years of post-Civil War peace marched on, fewer and fewer men had the benefit of wartime hardships in jump-starting this faculty:

> If the [Civil] War had done nothing else for our people, it would have done much simply in teaching them that deeds are greater than words. The American people, through those long days of anguish and suspense, learned how much higher and nobler is the power that can do and dare and endure, than the arts of dainty expression, or vehement declamation, or cunning dialectic in which they had formerly so much delighted . . .[28]

In the absence of countervailing combat experience, the traditional methods of learning in secondary education were positively dangerous to a young man's executive faculty:

> No one familiar with the laws of mind will be disposed to deny that there is at least a tendency, in the protracted study of any subject, apart from putting that study to a practical use, toward producing a partial paralysis of the will, shown in the disposition to procrastinate, to multiply distinctions, and to stand shivering on the brink of action.[29]

Another feature worth noting in Walker's outline for industrial education is the obtrusive visualism. Facility in visual observation and graphic representation were skills Walker considered crucial for both workers and elite scientists. He associated drawing skills closely with precision in both planning and execution of tasks, and more importantly, he saw graphic representation as an important means of forging links of understanding between the two main

---

[27] F. Walker, "Industrial education," address before the American Social Science Association, Sept. 9, 1884, reprinted in *DE*, 141–143.

[28] F. Walker, "The rise and importance of applied science in American education," address at the Convocation of the University of the State of New York, Albany, NY, Sept. 7, 1891, reprinted in *DE*, 31–32.

[29] F. Walker, "A plea for industrial education in the public schools," address before the Conference of Associated Charities of the City of Boston, Dec. 2, 1887, reprinted in *DE*, 157. It is of course possible to over-interpret, but Walker's use of the term "executive faculty" suggests just the sort of parallel between public government and self-government that Foucault was concerned to bring out in his writings on governmentality. It is not so far-fetched to imagine that Walker felt such a link at least vaguely, since his own life was characterized by an unusual degree of overlap between public and private identities, and since he identified quite strongly with the institutions of the Federal government.

strata of his ideal society. His work on the 1874 statistical atlas had convinced him that skillful graphic portrayal of information held great potential for promoting harmony between government and the governed. By making sometimes complicated statistical information accessible to properly prepared citizens, the scientists who would ideally inform government policy stood a much better chance of commanding general assent for their policy recommendations.

This brings us to another major benefit Walker saw in industrial education: social decency and reconciliation to the maintenance of social order. Walker had in mind here at least three things: the decent and respectful conduct of workers in their dealings with capitalists and employers; cheerful submission to the necessary importunities of the state; and the maintenance of an orderly public façade at the level of the household. With regard to the first of these, Walker hoped that education (in the most general sense of life experience as well as schooling) would lead workers to temper

the rightful and even necessary spirit of self-assertion . . . by so much of wisdom and self-control, by so much of a disposition towards fairness and reasonable concessions, as will lead them to seek their own good only through means which are compatible with the steady and even progress of production and with the due accumulation and conservatism of capital.[30]

Submission to state power was another goal, as is clear, for example, in his frustration with the need to tax citizens indirectly. In his view, an upright citizen should understand and accept the need for taxation. In a paper published in 1888, Walker looked forward to a time

when, through instruction of our children in civics, ethics and economics, and through the long-continued enjoyment of political franchises, governments shall be found, immediately subject to popular control, which shall yet be able to collect by direct assessment and exaction that tenth or fifth part of the laborer's wages which is now conveyed away from him by disguised imposts upon the decencies, comforts and luxuries of life.[31]

Finally, Walker hoped manual and industrial training would promote cleanliness and order in the publicly visible aspects of family life. This might have seemed a minor point, were it not for the heavy burden of explanation Walker placed on such matters in accounting for the decline of native White birthrates (see Chapter 6). It is not difficult to recognize the imagery he employed in extolling "the virtue which a general mechanical education of the people would have in preserving and exalting the priceless sense of social decency which keeps the fence along the village street in order, the gate hung, the glass

---

[30] F. Walker, "The laborer and his employer," lecture delivered at Cornell University, Feb. 1889, *Scientific American*, June 1, 1889, supplement no. 700, quote from typescript, Francis A. Walker Manuscript Collection, MIT Archives and Special Collections MC 298, Box 4, Folder 3, 21.   [31] Walker, "The bases of taxation," 86–87.

set, the shutter in place . . ."[32] Since the fate of all that was distinctive of American manhood, and perforce of American social health, hinged (if you will pardon the expression) on the state of repair in which shutters and gates were kept, it is not surprising that Walker returned to this image again and again. If these outward manifestations of social decency were unhung and accompanied by "green pools" in a yard crawling with naked immigrant urchins, the native white American man would "shrink" from the prospect of bringing forth the next generation.

If the primary purpose of public schooling was to produce an industrially mobile citizen-worker, appropriately civil in his dealings with employers, ready to face taxation, census takers and other government demands with equanimity, and able to keep up the public appearance of his home, Walker also had specific goals in mind with respect to young women. He devoted less attention by far to women in his writings on education, but he did hold them responsible for important aspects of the reproduction of labor and citizenry. If reproductive sex was entirely at the whim of the man, the Victorian-era domestic ideology so dear to Walker at least gave women an active role in food preparation, and keeping house. In his "Plea for industrial education in the public schools," Walker made it clear that the "industries" appropriate for young women were entirely domestic:

If, as Horace Mann said, it is a crime for a boy here to grow up in ignorance of reading and writing, what sort of offence is it, pray, for a girl here to grow up in ignorance of cooking and sewing? Think from what kinds of homes tens of thousands of our children in the public schools every morning come – rooms disordered and ill-kept, amid foul surroundings, presided over by a mother who cannot decently patch or darn a garment that is beginning to give way, and who knows only enough of cooking to take the perhaps abundant materials supplied her and render them, by dirty and wasteful processes, into disagreeable and indigestible messes, productive of dyspepsia and scrofula and provocative of a strong craving for drink. As a matter of public safety, can we afford to breed such a population in this Republic?[33]

Although it would be unrealistic to hold Walker to standards of perfect consistency across the whole breadth of his writings, and allowances must be made for changes of opinion, it is interesting to note that when it suited his purposes, he was willing to ascribe such things as alcoholism to immediate circumstances of "nurture" rather than long-bred "nature."[34] Walker's increasingly essentialist racism set strict limits on how much social benefit he could expect from educational reform, which could thus only form a part of his overall program for curing the ills of the American social body. But here at least, he could not fully despair of women's ability to raise children, since

---

[32] Walker, "Industrial education," 149.
[33] Walker, "A plea for industrial education in the public schools," 169–170.
[34] Compare endnote 1, 469–471, restored by D. R. Dewey to F. Walker, "The causes of poverty," *The Century* 55 (1897), 210–216, reprinted in *DES*, vol. II, 455–471.

to give up hope on this matter would be tantamount to releasing them for industrial labor outside the home. For Walker, legitimizing women's participation in the public workforce would dissolve the most important anchor of basic social order (see Chapters 5 and 6). The women he was comfortable training were implicitly *immigrant* women. In other respects as well (for example, sanitary regulation), no amount of pessimism *in principle* about the effects of regulation could quell the urge to regulate immigrants more aggressively than other groups. I will return to this issue below.

Public schooling would, Walker hoped, go a long way toward preparing a workforce of men fit for the struggles of modern times. But more was needed. At the end of the Civil War, "the industrial development of the country had reached the point where it had become necessary that the enterprises into which our labor and capital were to be put should be organized and directed with much more of skill and scientific knowledge than had been applied to our earlier efforts at manufactures and transportation."[35] To oversee and organize the efforts of citizen-workers, a new breed of citizen-scientists was urgently required. Science (particularly statistical social science) would also be of great use to the federal state, as Walker asserted countless times. But, in step with the general logic of governmentality (see Chapter 1), Walker preferred to have scientific education located outside state institutions. "[I]t does not follow, from the importance of science to the state, that science should be directly fostered or supported by government. It might conceivably be that science would do its work for the state better if the state itself did nothing for science . . ."[36] While one purpose of public education at lower levels was social integration of a national labor force and citizenry, an adequate level of integration and investment in social order could be assumed for college students. The more important task at the post-secondary level was to protect the *independence* (and hence objectivity) of scientific training and research. In this belief, Walker echoed Gilman, G. Stanley Hall, Andrew D. White and other advocates of the German university system.[37]

The curriculum at MIT was steadily shifted under Walker's guidance in the direction of an ever more practically oriented set of course groupings aimed at meeting concrete social and industrial needs. The grounding in basic sciences and mathematics remained mandatory, but Walker insisted that these serve primarily as the tools with which to approach hands-on problems. He took great pride, for example, in the establishment of a full course sequence in sanitary engineering, composed of instruction beyond the basic introductory level in chemistry, biology, sanitary and hydraulic engineering, and heating and ventilation.[38]

---

[35] Walker, "The rise and importance of applied science in American education," 19.
[36] Walker, "Socialism," 259.   [37] Rudolph, *The American college and university*, 245.
[38] F. Walker, *President's reports*, 1888–1889, 45; 1891–1892, 48, MIT Archives and Special Collections.

At the collegiate level as at the primary and secondary levels, Walker shared the educational program of his predecessor as president of MIT, John D. Runkle. Runkle was one of the most important early champions of "manual education" as a social adaptation to conditions in the post-Civil War industrial age. As the historian of education Lawrence Cremin tells it, Runkle and Professor Calvin M. Woodward of Washington University in St. Louis had been puzzling, during the mid-1870s, over how to translate their rough ideas on manual education into a workable curriculum. The watershed event for them was the Centennial Exposition at Philadelphia in 1876, more particularly a display mounted by Victor Della Vos, Director of the Moscow Imperial Technical School. Della Vos and his colleagues had themselves been searching for a way to integrate basic science and engineering courses with hands-on work. They hit upon what Cremin terms "the radically new pedagogical idea that one could analyze the skills required for each of the trades, organize them in order of ascending difficulty, and then teach them according to a program that combines drawings, models and tools by which a student could, under supervision, progress to a requisite standard of skills."[39] Della Vos had shops built for this portion of the curriculum which would also sell the goods produced by students. Commercial possibilities aside, Runkle and Woodward were most impressed, and felt that they had found the key to executing their program. Indeed, John W. Hoyt, who wrote the judges' summary on "Educational systems, methods and libraries" for the Centennial Commission, reported that the Russian exhibit was, in large part due to these displays, "in the estimation of many American and other educators, the most important representation made by any state or nation in this group."[40] Since Walker was chief of the Bureau of Awards at Philadelphia in 1876, he undoubtedly also knew of Della Vos's display, and in all likelihood met and conferred with Runkle as well. Like Runkle, he believed that the basic principles embodied in the Moscow Imperial Technical School curriculum should shape courses of study at all levels, and it becomes easier to understand why he was chosen to succeed Runkle.

Della Vos's curriculum is interesting as an early precursor to the time and motion studies later associated with Frederick Taylor and the movement for scientific management. It also suggests some of the reasoning behind Walker's stress on the importance of drawing and visual representation. In so far as the process of learning a trade ceases to rely on the presence of an accomplished

---

[39] L. Cremin, *American education*, vol. III, *The metropolitan experience, 1876–1980* (New York, 1988), 223–224.

[40] J. Hoyt, "Educational systems, methods, and libraries," in United States Centennial Commission, *International exhibition, 1876*, vol. IV. Reports and awards: Groups XXVIII–XXXVI (Washington DC, 1880), 161. See pp. 161–179 for a full description of the exhibit.

craftsperson or working operator (with all the knowledge that person would embody), and becomes instead a standardized sequence of increasingly complicated tasks, much of the formerly tacit spatial knowledge about shape, size and proportion which the operator would have made available must be rendered onto paper as blueprints and plans.

There is a link here with the training in graphic representation that formed a part of the statistics course inaugurated under Walker at MIT. In this course, students were taught "to construct diagrams and employ other modes of graphic representation, in application to a great variety of subjects, vital, industrial, commercial and political."[41] Walker's disciple Davis R. Dewey (who also edited the volumes of his collected "Discussions") taught it, though Walker wished he himself had the time to take charge. Years before, in 1876, when invited by Daniel Coit Gilman to spend a term at Johns Hopkins as a visiting professor, Walker had first proposed as a title for his course "Graphic illustrations of statistics." He subsequently changed the course completely to address the "currency question" which was then occupying Congress and the interested public, and afterwards turned his lecture notes into the well-received book on *Money*.[42] But graphic illustration remained dear to his heart, and integral to his conception of public communication. Whether in illustrating statistical information or in guiding a manufacturing process, accurate drawings, charts, diagrams and blueprints are technologies for improving the communication of expert knowledge to an as-yet untutored audience (whether students or the general citizenry). To stretch the analogy even further, these two types of visual representation share something important with the orderly urban street patterns which James Scott links with the modern planning mentality (see Chapter 5). A rectilinear urban landscape, too, allows the uninitiated to make their way through an unfamiliar process or terrain.

To return to Walker's educational program, as the presence of a statistics course suggests, Walker was not in the business merely of training modern, specialized scientific experts; he insisted that his charges receive some sort of broader education that would allow them to see how their scientific and technical expertise related to wider social issues. Thus, MIT students were required to take at least a bare minimum of courses in history and political thought. After all, for Walker, the main issue in educating scientists as in educating workers was still a traditonally conceived *manhood*.

In none of the higher walks of life does it ever cease to be more the question how much of a man one is, than how much he knows of his special business . . . A great lawyer

---

[41] F. Walker, *President's reports* 1887–1888, MIT Archives and Special Collections.
[42] F. Walker to D.C. Gilman, Feb. 14, 1876, Nov. 4, 1876, Daniel Coit Gilman Papers, Special Collections, Johns Hopkins University, MS 001.

generally is a great man, but he need not be: there is a melancholy abundance of instances to the contrary. But a great engineer must be a great man. All great engineers, according to the testimony of those who knew them, have been great men.[43]

It was this belief that underlay Walker's pride in the seriousness with which MIT students had to perform the military exercises then required of all state-supported institutions of higher learning, but carried out in a minimal and perfunctory fashion at most colleges. Military drill caused students to "acquire the instinct of subordination [to the demands of such disinterested pursuits as scientific research], the power of cohesion, promptitude and precision of movement, and habits of mind as well as of physical bearing which cannot fail to be useful in a high degree."[44]

Military drill also cultivated endurance, which was quickly coming to be known as an indispensable quality for MIT students. First in the Boston academic community and then throughout the eastern United States, MIT under Walker rapidly gained a reputation for working its students to or beyond the point of physical breakdown. In a memorial pamphlet published after Walker's death by the Massachusetts Commandery of the Military Order of the Loyal Legion of the United States, Colonel Thomas L. Livermore marveled at Walker's success in keeping his hard-pressed charges diligently at work: "The enthusiasm which President Walker imparted to his students was like that which the successful military leader inspires in his soldiers, and the cheerfulness with which these students, under his extraordinary influence, submitted themselves to the severest and most unremitting intellectual labor was a phenomenon in the history of education."[45] By all accounts, Walker was indeed a personal inspiration to many students, but the concern over the strenuous workload was real, widespread and public enough to prompt action. Characteristically, Walker decided to collect information on the matter, and much of the surviving correspondence he received as President during the mid-1880s takes the form of responses from former students to a circular questionnaire he sent out. Though I have been unable to locate a copy of this letter, it basically asked the question, "did you find the course of study at MIT to be so strenuous as to threaten your health?"

The responses are interesting in a variety of ways, not least in what their letterheads reveal about the employment patterns of MIT graduates during

---

[43] F. Walker, "The relation of professional and technical to general education," *Educational Review* (1894), reprinted in *DE*, 59.

[44] F. Walker, *President's reports*, Massachusetts Institute of Technology, 1882/83, MIT Archives and Special Collections, 25.

[45] T. Livermore, "In memoriam – Companion Brevet Brigadier General Francis Amasa Walker," Military Order of the Loyal Legion of the United States, Commandery of the State of Massachusetts, Circular no. 2, series 1897, in Francis A. Walker papers, MIT Archives and Special Collections, MC 298, Box 5.

the 1880s.[46] More to the point here, most respondents followed a standard script implicitly written by the prevailing culture of manhood. According to this script, the respondent acknowledges that the workload was a constant topic of conversation among students, and perhaps even that he himself (almost all respondents were men, reflecting the makeup of the student body) had had health troubles. But any such troubles are dismissed as the product of the respondent's own immaturity and lack of seriousness, or, if he had no such troubles, are ascribed to the immaturity of his trifling classmates. Some former students ventured the opinion that the workload left absolutely no time for anything else, but this proposition was usually followed with an assurance that the workload was no more strenuous than necessary. In short, almost all responses were just what Walker would have wanted to hear.

Nevertheless, the persistence of this public relations problem caused him to remain preoccupied with the issue into the 1890s, puzzling over how best to reconcile the need to work through a scientific curriculum with the need to keep his students vigorous and robust. In thinking about the different things students needed to spend time on, he realized that the daily commute to and from school might represent a solvable problem. In his President's Report for 1893–94, Walker published the results of an informal geographical study he had made, with the help of one of his faculty members, a Dr. Ripley. He divided 1,177 students into five classes based on time and difficulty of commute, finding that 641 lived within Boston and nearby Cambridge (and probably commuting less than 20 minutes each way), 215 in nearby suburbs (Brookline, further Cambridge) requiring 20–40 minute trips by streetcar, 181 between 1–10 miles distant on rail lines, 56 between 10–20 miles away and 84 further than 20 miles away. On the basis of Walker's data, Dr. Ripley estimated the overall "daily migration" to and from the institution at greater than 8,700 student-miles. Although Walker was not explicit about it in his Report, he was clearly building a case in his late years at MIT for the construction of on-campus dormitories. It is interesting that although the scientific men Walker was molding would be highly mobile after graduation (fanning out to all corners of the nation), the demands of their training as men were so severe as to require their temporary immobilization.

Women had been a presence at MIT since before Walker arrived, though their numbers remained small. In the early 1880s, the number of women on campus fluctuated in the teens and twenties out of a total student population

---

[46] A sampling: American Bell Telephone, Saint John Bolt and Nut Company, Office of the City Engineer (Lowell, MA), Blue Hill Meteorological Observatory, the Berkeley School (Providence, RI), Schuyler Electric Light Company (Hartford, CT), Keystone Bridge Company (Pittsburgh, PA), Abbott Worsted Mill, Tremont Nail Company, Boston City Hospital, Sprague Electric Railway and Motor Company, Hamilton Binder Company. In Francis A. Walker papers, MIT Archives and Special Collections, MC 298, Box 1, folders 5–9.

ranging from roughly 200 to roughly 400; by the time of Walker's death in 1897, the student body had grown in size to 1,594, but women numbered only 71. Walker had little to say specifically about female students at MIT, though it is possible to surmise from his writings for the general public that their presence must have been at least somewhat troubling. In an 1892 article for the *Educational review* on "Normal training in women's colleges," Walker hinted strongly at his prejudices in his introductory comments:

I shall not pause to inquire whether this object [fully equal four-year education for women] was in itself desirable; whether young women should be called upon to do in four years all that young men may be required to do in the same time; whether exceptional consideration be not due to the greater delicacy, sensitiveness, and liability to nervous derangement on the part of the female sex. I shall not even stop to ask whether this claim [i.e., that women are up to fully equal schooling] was ever, anywhere, made good except in the case of highly selected bodies of young women; or whether, through conscious or unconscious relaxations of the nominal requirements, the work exacted, in even the most advanced women's colleges, has not, in fact, mercifully fallen somewhat below a full equivalent of that done in the other class of institutions.[47]

In the argument that follows this none-too-subtle opening salvo, Walker claims that most women who insist on going to college should be educated in general culture only, and not led into specialization, "first, because the presumption is still happily in favor of the ultimate devotion of woman's powers and faculties to domestic life and duties, in which general training will count for much, and special training but little," and second, because the businesses for which women *are* suited only require "woman's tact, dexterity, and quickness of apprehension," qualities best cultivated *generally*. The one exception to this recommended policy is teacher ("normal") training, since the profession of teaching "is mainly relinquished, by general consent, to women."[48] This was Walker's way of reconciling the conflicting demands of a system in which women were, on the one hand, judged constitutionally unable to participate fully in public life, and on the other hand, charged with preparing boys for that life.[49]

Walker's reflex in dealing with female students at MIT was to segregate them, both literally and figuratively. At the epistemological level, his Presidential Reports always included a separate section (usually a single paragraph) giving basic information on women students: how many there were, and what courses of study they pursued. For most of his tenure at MIT, the majority of female students were listed as "special" students, which exempted the school from having to include them in its official student roster. Walker was not alone in remaining uncomfortable with their presence. Physically, as well,

---

[47] F. Walker, "Normal training in women's colleges," *Educational Review* (1892), reprinted in *DE*, 306–307.   [48] *Ibid.*, 313–315.

[49] See S. Coontz, *The social origins of private life: A history of American families, 1600–1900* (New York, 1988) on the larger cultural significance of this contradiction.

women were set aside. From 1876 (before Walker arrived), the Institute had had a Woman's Laboratory affiliated with it, largely due to the efforts of one of the more interesting presences at MIT, Ellen Richards.[50] Richards had been MIT's first female graduate, and her Woman's Laboratory was a success, but she was not officially a part of the faculty. A year after Walker arrived, Richards and a handful of her students and assistants from the Woman's Laboratory persuaded Walker to incorporate into the plans for a new chemistry laboratory building a women's bathroom. The negotiations also resulted in the establishment of a separate women's reading room, named for Margaret Cheney, one of Richards' promising young students who died suddenly while the new building was under construction.[51] But Richards was not at first given a teaching position from which to continue her pedagogical work. Finally in 1884, she was appointed instructor in Sanitary Chemistry, a position she held until her death in 1911.

Walker characterized Ellen Richards as a mentor for the women students, and her biographer concurs, claiming she was the *de facto* dean of women at MIT. She ministered to her female charges in all ways, psychologically, financially and socially.[52] During the 1880s, Richards was already developing the ideas with which she would lead the "domestic science" movement in the early twentieth century. This movement was aimed, under Richards' direction, at training women to become full-fledged scientists. Richards envisioned women conducting experiments on foods, fabrics and cleaning materials in the home, and using the results of their inquiries to rationalize and optimize the processes of family management and child-rearing. Ironically (in light of Walker's views) she also intended that women use such skills in realms beyond the household, and enthusiastically encouraged women to get involved in "municipal housekeeping."[53] For his part, Walker was positively disposed toward the sanitary chemistry course, happily relating in his last President's Report that it enabled students to "gain much more than the simple practice necessary to fit them to carry out analyses of water, air, butter, milk, cereals, etc."[54] Perhaps he was not aware just how much "more" Richards hoped her students would do. Despite the misgivings of Walker and many other men, once inside the hallowed halls of science, women, too, began to put their training to use in the wider field of public policy.

I have organized this brief survey of Walker's educational program in order to stress its central concern with cultivating manhood. One way to sum it up is to say that while his program for primary and secondary education was

---

[50] C. Hunt, *The life of Ellen H. Richards* (Boston, 1912), 136. [51] *Ibid.*, 148–149.

[52] *Ibid.*, 151.

[53] S. Stage, "Ellen Richards and the social significance of the Home Economics movement," in S. Stage and V. Vincent (eds.), *Rethinking home economics: Women and the history of a profession* (Ithaca, NY, 1997), 17–33.

[54] F. Walker, *President's reports*, 1896–1897, 64, MIT Archives and Special Collections.

intended to replace the masculine training of a lost farm life, Walker organized education at MIT more to replace the masculine lessons of war, or at least basic military training. The accent was on endurance and strenuousness, and it was implicitly expected that the lives of the elite scientists and engineers who would go on after graduation to direct the progress of the nation should approximate Walker's own in their level of unstinting effort. But, however successful, the combination of manual/industrial education at lower levels and technical/scientific education in colleges, universities and technical schools would not completely address the threats to the integrity of the American social body. For American manhood was imperiled not only by the squandering of its productive potential in outdated educational practices but also (and more fundamentally) by the suppression of both its productive and its reproductive potential as a result of unhindered immigration. Education was all to the good, but if the human material on which it had to work was being continuously degraded by inferior racial stock, it would be a losing battle. In the largest sense, and in keeping with Walker's social Lamarckianism (see Chapter 6), the adaptability of the American social body through such measures as public education was circumscribed by race. Accordingly, the second main pillar of Walker's overall program for governmental regulation was more restrictive than fostering. The next section of this chapter will clarify the spatial logic by which Walker had to operate, and will recount his views and actions with regard to American Indians during his stint as Commissioner of Indian Affairs. This will make clearer by contrast the spatial features of the policy he pursued with regard to immigrants. I will end with a brief account of how his ideas on immigration restriction actually affected the important restrictive legislation of the early twentieth century.

The logic and historical results of spatial restriction

Regulation of a national body politic is inherently spatial in so far as the ends of regulation involve issues of who or what should be where. When social "health" is understood in terms of appropriate locations, appropriate degrees of mobility and fixity, or appropriate degrees of separation for people and things, concrete policies will incorporate such spatial devices as impermeable borders, segregation and sedentarization, on the one hand, and support for transport and communication infrastructures on the other. Coercive regulation will tend to focus on barriers to movement, while fostering regulation will tend to focus on enhancing mobility (though in cases such as enforced removal, coercion makes use of mobility). In addition, many regulations concerned with altering the behavior of individuals and families have a spatial component in so far as they involve invasions of nominally "private" space by "public" authorities. Francis Walker's understanding of the late nineteenth-century American social body clearly had spatial implications for regulatory policy. Not only did much of his construction and interpretation of census results revolve around questions of race and location; his explicit political economy also gave great significance to the mobility of labor. Further, his horror over the demographic prospects for native white Americans charged the border between public and private space with heightened significance.

Walker's desire to fix and restrict harmful or "inferior" elements of the national population was underwritten, if not entirely constructed, by the reifying effects of the census statistics in which he placed such faith. As noted in the section on "grids of specification" in Chapter 2, the very structure of census inquiries tended (as it still does) to encourage the idea that different groups of people somehow "naturally" belong in particular regions or locations. Walker inherited an emphasis on "race" that had characterized the United States Census since its inception in the tripartite constitutional division of the population into black, white and Indian categories. But Walker took this emphasis further, literally creating "foreigners" as a statistical "race" which he usually treated on a par with African-Americans or Asian-Americans. In his 1874 *Statistical Atlas*, for example, he treated foreigners, regardless of their countries of origin, as a single group with a characteristic pattern of susceptibility to different diseases (see Chapter 5). If a census taken before statistical correlations were fully understood tended to naturalize associations between significant demographic groups and regions or locations, and those groups were defined primarily in terms of race, some sort of racial geographic determinism was a predictable result. Thus, the diagrams Walker published with the results of the 1880 Census, which compared African-Americans and "foreigners" in terms of the latitudes, levels of annual rainfall, altitudes above sea level, and average annual temperatures most congenial to each group, make perfect sense (see Figure 10, p. 159). As noted in Chapter 6, Walker implicitly believed that there was a sliding scale of geographical adaptability, according to which the most vigorous races (those of Northern European stock) were most adaptable to different conditions, while the inferior races were more seriously constrained to stay within certain environmentally appropriate limits. Although he had fought for the Union in the Civil War, like most northern whites, Walker was far from believing that blacks were equal in their abilities to other races. Thus, in portraying "foreigners" in the same way as African-Americans, Walker was suggesting in yet another way that the former as well as the latter were inferior. Again, the problem in spatial terms was that these inferior and hence immobile foreigners were camped in the industrial heart of the nation, the northeast. Whatever Walker's solution to this problem, it would be motivated at least in part by the conviction that these immigrants were "out of place." Walker had plenty of experience with racially based exclusion and segregation during his tenure in the early 1870s as Commissioner of Indian Affairs. I will briefly recount the policy conclusions he drew from his stint at the BIA to give a sense of other spatial tools Walker decided not to advocate with regard to immigrants. This will help to highlight, through contrast, the circumspection and governmental prudence represented by his advocacy of immigration restriction.

Contrary to a suggestion of Ronald Takaki's, Walker was not the "father of the reservation system."[55] Reservation policy as it was when Walker took over the Bureau of Indian Affairs in 1871 had evolved slowly and fitfully from the Indian policy of the 1820s.[56] Walker's contributions to the management of Indian affairs were modestly original only in the realm of bureaucratic rationalization, where he took unprecedented steps to render the far-flung network of Indian agents more accountable to the central office and to the public. Nevertheless, as with most of his other public roles, he soon felt qualified to write what turned out to be a highly influential essay on "the Indian Question" for the *North American Review*. This essay concisely stated his understanding of Indian affairs as they stood at the time, as well as his ideas about what was to be done. In it we can find a succinct explanation of the purposes and advantages of segregation as a mode of social control.

For Walker, the "Indian question" divided itself into two parts, and the failure to recognize the distinction had been at the root of most of the acrimonious and long-running debate between (usually eastern) advocates of "civilization" and (usually western) advocates of a "war" policy. The first part of the question was, "what to do with Indians in so far as they are obstacles to the advance of white settlement and the construction of railroads?" The second part concerned "what to do with them in so far as they ceased to be obstacles?"[57] The problem with most white commentators was that they tended to lump all Indians into one or the other of these two categories, whereas, in reality, some Indians were still significant obstacles, and many were not. The remainder of the article outlines Walker's proposals for how to deal with these two distinct sets of Indians (not always neatly mappable onto different tribes). With regard to the "hostiles," whose numbers Walker estimated at 60,000, and who were concentrated mostly in the Plains region, he recommended a pragmatic but undignified policy of continuing to indulge their illusions of strength and meeting most of their demands for payments and subsistence while simply trying to keep them on reservations while white settlement and railroad building hastened the day that they would cease to be a threat. The arrogance of Walker's rhetoric makes clear that he considered that day inevitable:

Grant that some petty Sioux chief believes that the government of the United States feeds him and his lazy followers out of fear, or out of respect for his greatness: what

---

[55] See R. Takaki, *A different mirror: A history of multicultural America* (Boston, 1993), ch. 9.
[56] See F. Prucha, *The great father*, 2 vols. (Lincoln, NE, 1984) for a comprehensive history of US policy toward Native Americans.
[57] F. Walker, "The Indian question," *North American Review* 66, 239 (1873), 337.

then? It will not be long before the agent of the government will be pointing out the particular row of potatoes which his Majesty must hoe before his Majesty can dine.[58]

While Walker believed the US Army could vanquish such groups as the Lakota ("Sioux") at any time, he considered the inevitable attendant sacrifice of white settler lives, and the comparatively vast expense of a war covering so much territory, prohibitive. Better, in his view, to suffer minor indignities while the passage of time brought an ever more complete net of railways and white communities to hem in and immobilize the "savages."[59] Walker had a keen understanding of the importance of reservations, in the interim, in allowing the government to punish the worst outrages of the "hostiles" while minimizing the likelihood of frontier war. During the 1870s, Indian agents throughout the Plains informed the tribes of the policy (which did not originate with Walker) that all Indians found outside reservation boundaries without permission were automatically defined as "hostile" and thus vulnerable to pursuit and violence from the US Army, while all observing a minimal orderliness and remaining within the reservations would enjoy the great advantages of "civilized" treatment by the government. Walker understood that because of this neat method of spatialized categories of Indians, "military operations thus conducted are not in the nature of war, but of discipline, and are so recognized by the tribes whose marauding bands and parties are scourged back to the reservations by the troops."[60]

Interesting though this part of his policy is, it was the other part, regarding Indians who were no longer threats to white settlement or the advance of railroads, that provides the most intriguing model for how Walker might have wished to deal spatially with immigrants. The situation of immigrants in relation to white society was of course very different from that of Indians, and the differences were decisive in explaining why Walker's policies toward the two groups differed. But leaving these differences aside for the moment, it is instructive to dwell briefly on the extent to which Indians and immigrants posed similar problems for the American social body.

Both groups posed serious threats as *reservoirs of inferior blood.* In Chapter 6 I outlined Walker's racism and his fixation on immigrants as sources of "pollution." In his discussion of Indians, the same concern looms large. Walker ends his paper on Indian policy with a long meditation on the anticipated results of dissolving Indian reservations and thus allowing the dispersal of Indians throughout white America. As far as whites were concerned, such a policy would bring a "debasement in blood and manners," a

"burden of vice, disease, pauperism and crime upon a score of new states more intolerable than perpetual alarms or unintermitted war; while the ultimate result of thus dispersing the Indian tribes among the settlements would be to multiply threefold

[58] *Ibid.*, 348.   [59] *Ibid.*, 345–352.   [60] *Ibid.*, 355.

within a century the number of persons having Indian blood in their veins. Surely this is not the way in which we wish to see the Indian problem solved! When one considers by what men and women, and with what patience, soberness, and faith, the foundations of the now great States of the Northwest were laid, he can but contemplate with dismay the prospect of a new generation of States of which ranchmen and miners are to be the fathers, and Indian squaws the mothers.

But if, on the other hand, the policy of seclusion shall be definitely established by law and rigidly maintained, the Indians will meet their fate, whatever it may be, substantially as a whole and pure race.[61]

It is difficult to determine how much more threatening Walker considered Indian than immigrant blood *per capita*. But if it can be assumed (from his immersion in a racial discourse that generally placed European races above Native Americans) that a given quantum of Indian blood was a graver threat, it is also true that there was far more European peasant blood (in absolute terms) moving about through the white population. And for Walker, the problem of degraded blood, from whatever source, only grew more crucial as time went by. Thus, it is not unreasonable to imagine him being tempted by quarantine measures and policies of seclusion of immigrants. His prescriptions for Indian "civilization" would have made some sense in the new context, and would have given freer rein to the "exclusionary mentality" identified by Sibley (see introduction to Part II above).

The key to Walker's recommendations for dealing with Indians who were no longer a military threat was, again, the reservation system, but he urged a more rigorous attention to its spatial logic. First, he recommended that every effort be made to render the reservations as isolated from white society as possible, so as to miminize "intercourse" in all senses, especially with the less scrupulous elements of white society. To this end, he also urged that reservations be shaped so as to be as spatially compact as possible and so as to offer the minimum boundary length per unit of area enclosed. For the same purpose, the number of reservations should be reduced and the Indians consolidated into one or two grand reserves or colonies. The boundaries of reservations should be "hardened" by elevating the existing administrative prohibition against whites entering without permission to the status of law, and enforcing it more strongly.[62] Similarly, the prohibition against Indians leaving without permission should be written into law. "Nothing but the knowledge that he must stay on his reservation, and do all that is there prescribed for him . . . will . . . prevent a general breaking up of Indian communities and the formation of Indian gypsy-camps all over the frontier States and Territories, to be sores upon the public body, and an intolerable affliction to the future society of those communities."[63] Further, Walker recommended that a "rigid reformatory control should be exercised by the government over the lives and manners

[61] *Ibid.*, 383, 384–385.   [62] *Ibid.*, 365, 368, 372–373.   [63] *Ibid.*, 375.

of the Indians . . . particularly in the direction of requiring them to learn and practise the arts of industry."[64]

The justification Walker gave for this last measure is especially interesting because it is clearly anticipates some of his later heated rhetoric regarding immigrants:

The right of the government to exact, in this particular, all that the good of the Indian and the good of the general [white] community may require, is not to be questioned. The same supreme law of public safety which to-day governs the condition of eighty thousand paupers and forty thousand criminals within the States of the Union, afford ample authority and justification for the most extreme and decided measures which may be adjudged necessary to save this race from itself, and the country from the intolerable burden of pauperism and crime which the race, if left to itself, will certainly inflict upon a score of future States.[65]

There were two major reasons why Walker did not in the end follow his restrictive impulses and recommend such draconian spatial regulation of immigrants within the territory of the United States, however tempting such regulations might have seemed in light of the problem of "blood." These reasons, of course, have to do with important differences between Indians and immigrants in their relations to white American society. The first, and perhaps the most important, major difference between the two groups was in political status: immigrants were expected to become citizens and thus acquire political power; Indians were members of "domestic dependent nations," or wards of the state. Citizenship was only anticipated for significant numbers of them far in the future. Thus, like colonized peoples elsewhere, they could be subjected to more aggressive policy manipulations than could immigrants. By contrast, it would have struck many of Walker's white contemporaries as outrageous to propose forcible segregation of immigrants, however "backward" they might have seemed.

The second major difference was in economic status. Unlike immigrants, Indians were by the 1870s largely irrelevant to the national economy, except in so far as they still occupied land that could have been turned to productive agriculture. This is an important difference, because it meant that the US government had no "governmental" motivation for taking a hands-off approach to Indian societies: there was no need to respect Indian lifeways as integral parts of an independent economic system on which the white economy relied. But immigrants were needed as laborers in the industrial economy. To restrict the movement of immigrants in order to subject them to an intensive "civilization" program would have meant violating one of the first principles of Walker's political economy: that *the mobility of labor is a key to avoiding class struggle*. However immobile immigrants might already have been, compared with the vigorous native whites, restricting them further

[64] *Ibid.*   [65] *Ibid.*, 376.

would only reproduce the horrors of the English settlement laws which Walker had so vehemently scourged in his treatise on wages (see Chapter 6). The need to honor and foster mobility was one of Walker's most sacred governmental principles.

This, then, was the fundamental dilemma facing Walker: he had many reasons to want to restrict immigrants already in the country, but was prevented by the Bill of Rights and by his own governmental principles from doing so. Once in the country, immigrants inhabited a national economic territory which had to be kept open for whatever movements of people, resources or goods economic forces might require. Adult men had to be free to move about. Thus, Walker's regulatory policy could only address the various borders of this national public space. His educational policy was intended to prepare young men before they crossed the border into adulthood. He also supported (though he did not write much about) the urban sanitary regulations that invaded the private homes of immigrants, not only in principle, but also through the "sanitary engineering" curriculum at MIT. Just as childhood was a liminal time, the home was a liminal space in relation to the broader socioeconomic public space of adult men, a space simultaneously *other* than public space in its organization and *necessary* for public space to exist at all. Because it was necessary for the functioning of economic life and yet not subject to the strictures of governmental caution, private space, too, could be considered fair game for regulation. Sanitary regulations were in Walker's view fully within the basic police powers of the state:

I do not mean to say that I should hesitate to approve of sanitary inspection and regulation, carried to their extreme, if they were as socialistic as anything ever dreamed of by Marx or Lasalle. For such good as I see coming from this source, in the reduction of vicious instincts and appetites, in the purification of the blood of the race, in the elimination of disease, I would, were it needful, join one of Fourier's "phalanxes," go to the barricades with Louis Blanc, or be sworn into a nihilistic circle. But . . . [t]he protection of the common air and the common water comes within the police powers of the states by no forced construction, by no doubtful analogy.[66]

But education and sanitary regulation assumed that the legacy of inherited blood was mutable, and, more importantly, that corrective measures could keep pace with the arrival of immigrants. Again, Walker became more pessimistic on both counts as the 1880s gave way to the 1890s. Thus, native white working men would be forced to continue to move through a socioeconomic space littered with unsightly warrens of unwashed degenerates. Since most immigrant ghettoes were located in and around the industrial districts through and among which men had to circulate in pursuit of economic advantage, there was little hope that native white men would be spared the spectacle so

[66] Walker, "Socialism," 256–257.

injurious to their reproductive ardor. The "fixity" represented by patches of undesirable blood in the midst of the field of "mobility" was unacceptable. As a result of this inexorable logic, immigration restriction began to assume fundamental importance for Walker as the *sine qua non* of any attempt to protect the native white American manhood that constituted the lifeblood circulating in national economic space. The erection of borders *within* the national territory too clearly violated cherished political economic as well as democratic principles. Thus restrictive regulation would only make sense at the outer boundaries of the American social body, as a spatial prophylaxis.

Walker did not come out openly in favor of immigration restriction until the early 1890s, after he had thoroughly convinced himself that immigration was the source of most of the country's pressing problems. His own proposals seem relatively modest in retrospect, but would have appeared much more radical at a time when the United States had as yet known nothing of restriction of Europeans. In an 1892 article for the *Yale Review*, after considering a variety of alternative schemes, he proposed that each immigrant be required to deposit $100.00 with the government upon entering the country. This amount was to be refunded upon exit to immigrants who left within three years, or, for those who intended to stay, would be refunded after the same period of time "upon the presentation of satisfactory evidence that [they are] at the time . . . law-abiding and self-supporting citizen[s]."[67] This law "would not prevent tens of thousands of thrifty Swedes, Norwegians, Germans, and men of other nationalities coming hither at their own charges, since great numbers of these people now bring more than that amount of money with them."[68] But it would prevent the "pipeline" immigration organized by steamship companies, which decanted the poorest Southern and Eastern European peasants onto American soil.

However, Walker's specific proposals as to how to accomplish restriction were less influential than his argument about the threat posed by immigrants to native white "stock." A group of young Harvard graduates centered around Charles Warren, Robert DeCourcy Ward and Prescott F. Hall had found in Walker's dramatic spectacle of native white men "shrinking" from reproduction a basis for action, and in 1894, formed the Immigration Restriction League. Walker's theory had "satisfied the incipient desire to justify the end of the free immigration policy," had "bridged the gap between the Brahmins' new inclination to stop the flow of European immigration, and their economic and democratic inhibitions against the idea of restriction."[69] Walker was offered

---

[67] F. Walker, "Methods of restricting immigration," *Yale Review* 1 (1892), 138–143, reprinted (in part) in *DES*, vol. II, 430.   [68] *Ibid.*

[69] B. Solomon, *Ancestors and immigrants* (Cambridge, MA, 1956), 99–103, 76, 69. John Higham, *Strangers in the land: Patterns of American nativism, 1860–1925* (New Brunswick, NJ, 1955), 142–149, gives Walker a similarly prominent place in the story of early restriction agitation.

the presidency of the IRL, but (like other sympathetic and prominent New England men such as Henry Cabot Lodge), he declined, both because he was so busy and because he retained a prudent awareness that restriction was as yet an unpopular idea among the general citizenry.[70]

Walker's "unhung gate" theory of decline in native white population growth began to come under attack as early as 1893, but, as Jean-Guy Prévost argues, it continued to shape debate about immigration into the 1910s.[71] In the years around the turn of the century, it was "more and more widely discussed, with hardly anyone equipped statistically to challenge it."[72] To put it in modern terms, statisticians and social scientists had not yet quite gotten their minds around the problem of spurious correlation. The particular conceptual elements of Walker's thinking which had the most lasting effect were his notion of "pipeline" immigration and his early attempts to suggest that the distinction between "old stock" and "new stock" immigrants could be made scientific through the use of statistics. Whereas earlier advocates of free immigration had occasionally relied on the social Darwinist argument that immigrants were self-selected carriers of unusual vigor, Walker "neatly turned the tables, declaring that natural selection was now working in reverse. Due to the cheapness and ease of steamship transportation, the fittest now stay at home; the unfit migrate."[73] With regard to providing a statistical basis for the old-stock versus new-stock distinction, Walker had had two contributions to make: first, in setting a general example with his efforts to reify "foreigners" as a statistical category; and second, in having made sure that information about foreign birth and foreign parentage were collected in 1870 (despite the lack of specific provision for it in the enabling legislation) as well as in 1880. In other words, he produced something more than a "gestural mathematics" in support of his governmental program.[74]

It remained for one of Walker's intellectual inheritors at the newly permanent Census Office, Clinton Rossiter, to refine Walker's statistical reification so as to encompass the distinction between old and new stocks. In *A century of population growth, 1790–1900*, which came out in 1909, Rossiter reported numerical estimates of the proportion of the American population descended from the people enumerated in 1800. His calculations showed that 35 million of the 67 million people counted in 1890 were of old stock. As Margo Conk puts it, "[h]ere at last was a positive and precise measure of the contribution of the "native element" to the American population – a number that could be

[70] *Ibid.*, 104.
[71] J.-G. Prevost, "Francis Walker's theory of immigration and the birth rate: An early twentieth-century controversy," Cahier no. 9701, Cahiers d'épistémologie, Groupe de Recherche en Épistémologie (Montreal, Quebec, 1997), 7–8.   [72] Higham, *Strangers in the land*, 147.
[73] *Ibid.*, 143.
[74] M. Poovey, *A history of the modern fact: Problems of knowledge in the sciences of wealth and society* (Chicago 1998), 172.

used to quantify just how far an unlimited and indiscriminate immigration policy had taken America from its Republican roots."[75]

Rossiter was one of a triumvirate of Census Office employees in the early twentieth century (the other two being Assistant Director Joseph Hill and Director Simon Newton Dexter North) whom Margo Anderson credits with keeping Walker's ideas at the center of the immigration debate. In publications both official and unofficial, these three hammered away at free immigration, supporting, if not actually coordinating with, the IRL. For example, SND North's long essay on "Seventy five years of progress in statistics," given at the 75th anniversary of the American Statistical Association, included a strong plug for restriction. He warned that, "[s]hould immigration continue on its present scale, should the disparity in the fertility of foreign and native stocks also continue, our population, which at the time this Association was founded was almost wholly Anglo-American, and in 1900 half native and half foreign, may in 1950 be three-fourths or more of foreign blood."[76]

The 1911 reports of the Immigration Commission established by a worried Theodore Roosevelt also contain various traces of Walker's influence. The statement the Commission had received from the Immigration Restriction League echoed Walker on a number of points, claiming that "the same arguments which induce us to segregate criminals and feeble-minded and thus prevent their breeding apply to excluding from our borders individuals whose multiplying here is likely to lower the average of our people." Immigrants were undeniably "lower," since "[a] considerable proportion of [them] now coming are from places and countries, or parts of countries, which have not progressed, but have been backward, downtrodden, and relatively useless for centuries."[77] In response to proposals that immigrants be subject to mandatory dispersal throughout the national territory, IRL President Prescott F. Hall held up recent "scientific findings" that "heredity is a far more important factor in the progress of any species than environment."[78] Thus,

[c]hange of location from Eastern States to other States will not change the character and tendencies of an immigrant even as much as the change from Europe. Unless the distribution is very wisely done, and possibly even then, the result will be the spreading of big slums over the country in the form of little slums . . . [Indeed, distribution from large seaports would act] as a force pump [vacuum] to draw in even larger numbers . . . The steamship companies would, no doubt, be glad of a vacuum of this

---

[75] M. Conk, "The Census, political power, and social change: The significance of population growth in American history," *Social Science History* 8 (1984), 94.

[76] S. N. D. North, "Seventy five years of progress in statistics: The outlook for the future," in J. Koren (ed.), *The history of statistics: Their development and progress in many countries* (New York, 1918), 41.

[77] US Immigration Commission, "Statement of the Immigration Restriction League," in *Statements and recommendations submitted by societies and organizations interested in the subject of immigration*, Senate Document 764 (61st C., 3rd S.), US Serial Set 5881 (Washington, DC, 1911), 106–107.   [78] *Ibid.*, 106.

kind, and would see that the "pipe line," as Gen. F. A. Walker called it, was constantly in operation.[79]

The parallel between this passage and Walker's earlier discussion of the results of dispersing Indians away from reservations is striking. The IRL was more bold than Walker had been in contemplating spatial manipulation, and also in conceiving of immigration restriction as a form of spatial eugenics. However, just before his death, Walker, too, had taken a transparently eugenicist line in linking pauperism to immigrant blood:

We must strain out of the blood of the race more of the taint inherited from a bad and vicious past before we can eliminate poverty, much more pauperism, from our social life. The scientific treatment which is applied to physical disease must be extended to mental and moral disease, and a wholesome surgery and cautery must be enforced by the whole power of the state for the good of all.[80]

The Immigration Commission accepted many of Walker's ideas, most importantly the notion that "the old and the new immigration differ in many essentials," and acknowledged the importance of his collection of foreign parentage statistics starting in 1870, and place of birth of foreign parents in 1880 as key bases for subsequent "scientific treatment" of the immigration issue.[81] The Commission's reports would continue to serve as a reference source for officials and advocates through the 1910s and into the 1920s.

   The movement to establish strong restrictions on immigration faced serious difficulties through the 1910s, as foreign countries made it clear that they would not take kindly to any quota system that favored some regions or countries over others. In the end, Rossiter's statistically updated version of Walker's old stock versus new stock distinction resolved the matter. The 1924 legislation that represented the culmination of the restriction movement justified differential quotas as a mechanism for preserving the American "national character," which was assumed to be determined by the proportions of different national stocks making up the population. The legislation used the proportion of old and new stock descendents at an earlier census as a baseline, and set national quotas as fixed percentages of the number of Americans of corresponding descent residing in the country at that earlier time. At first, the 1910 Census had been proposed as a baseline, but restrictionists argued that it would result in too generous quotas for Southern and Eastern Europeans. After much wrangling, 1890 was accepted as the baseline to be "preserved," and the famous law of 1924 passed.[82]

[79] *Ibid.*, 109.
[80] F. Walker, "The causes of poverty," *The Century* 55 (1897), 210–216, reprinted in *DES*, vol. II, 469.
[81] US Immigration Commission, *Abstracts of reports of the Immigration Commission*, Senate Document 747 (61st C., 3rd S.), US Serial Set 5865 (Washington, DC, 1911), 13, 122.
[82] M. Anderson, *The American census: A social history* (New Haven, CT, 1988),140–150.

This chain of events was arguably Walker's most important legacy to national policy-making, and it is not surprising that scholars of immigration restriction consider him a key founding figure. It also represents better than any of his other accomplishments the culmination of his program of governmentality. Not only is it a consistent solution, in general terms, to the problems he saw when he contemplated the American social body, but more specifically, it is a fitting resolution of his multilayered quest to master the national territory through a modern spatial politics. Territorial mastery has an epistemological dimension (involving abstraction, assortment and centralization) as well as a regulatory dimension (involving establishment of the right to regulate, as well as more obviously spatial matters such as the ability to confine and segregate elements of the population, and to invade private space in the "public" interest, or the ability to protect industry). Walker in effect concluded that the condition of possibility that needed to be in place for all these other aspects of mastery to be effective was the ability to control the flow of people through the borders of the territory. Twenty-seven years after his death, the clear definition he had given to the western edge of the American social body (see Chapter 5) was finally complemented with a meaningful eastern limit. The Atlantic shore had of course always been at least as clear as the western white frontier in *epistemological* terms, but the flow of people across it was more significant as a threat to the integrity of the social body. In the West, where white America was expanding, there was no need to fortify the epistemological boundary with physical control of human movement. Transgression of the boundary by whites was a sign of health, and led to the "ingestion" of more territory. In the East, by contrast, the social body was being "invaded" and penetrated by populations capable of challenging the nation's "proper" racial composition. The boundary of the social body had to be strengthened and the flows regulated.

Part II of this book has been concerned to show how different kinds of spatial politics have infused every moment of Walker's program, from observation through normalizing judgment to regulation. At the same time, I have tried to demonstrate the connections between a geographical construction of the American social body (Chapter 5) and such geographical regulatory strategies as immigration restriction, keeping the focus as much as possible on the spatial logics at work. Riding piggyback on Francis Walker as he moved through an astonishingly wide array of public positions has allowed me to find many contact points between the logic of governmentality and the various more familiar concerns of the late nineteenth century, without taking on too vast a subject.

Conclusion

In this final chapter, I will draw together the most important threads I have followed in different parts of this study. It is neither possible nor desirable to attempt to bring *all* of the empirical and theoretical issues to closure. Again, my overarching purpose has been to open up questions and draw new connections, not to settle matters of interpretation once and for all. Nevertheless, there are a few key things I especially hoped to accomplish with the story told here, related, first, to the place of theory in historical geography, second, to the development of the concept of governmentality, and third, to the historiography of the late nineteenth-century United States. In what follows, I will take these central points in order, with reference to the chapters that address them.

"Theory" and historical geography

According to the genre standards normally in force in historical geographic writing, this study is heavily laden with "overt" social theory. The extended discussions of governmentality in Chapter 1 (as well as in later chapters), the organization of Part I according to an archaeological framework taken from Foucault, and of Part II according to what I have called the "cycle of social control," might well seem obtrusive. One of the important purposes of this study has been to challenge the intellectual aesthetic according to which theory should be kept to a minimum, to challenge the idea that overt "regulation" of one's narrative by pregiven conceptual structures is somehow worse than the attempt to "let the facts speak for themselves." Much in the way that Francis Walker and others questioned the economic principle of *laissez-faire* as a guide for practical policy, I and many others before me have begun to question the doctrine that facts can speak for themselves as a guide to the construction of an historical argument. If we cannot avoid responsibility for structuring facts, the attempt to be covert and unobtrusive about it holds great potential for misleading readers.

The aspect of this study that represents the most obtrusive incursion of theory where it would not be expected is the use of archaeological method in Part I. It is not suprising to find the discussion of governmentality in Part II structured by theoretical questions, but shaping even the preliminary survey of historical context in such an overtly theoretical way may seem excessive. Thus, I would like to dwell once more on the point of the archaeological narrative. In Chapter 1, following the suggestion of Nikolas Rose and Peter Miller, I characterize the historical emergence of the logic of governmentality as in the first instance a discursive phenomenon. This logic appears initially in the guise of a set of programs for state action, programs proposed by advocates orally or in print, but in most cases destined for partial implementation at best. Foucault's archaeological method, in particular his notion of a "discursive formation," is well suited as a template for describing governmentality in its concrete emergence. One reason is that I have been concerned from the beginning to highlight the *constructed* nature of the American social body. The analytical distinctions Foucault makes under the headings of "formation of objects" (see Chapter 2) and "formation of enunciative modalities" (see Chapter 3) are peculiarly apt: they help show how the object and the subject proper to the discourse of governmentality cannot be separated in their origins from the specific conditions of emergence of the public discussions in which they first appeared. In the absence of this basic insight, it might have been far easier to assume that such categories as "the American social body" and "government experts" were "natural" and unproblematic. This assumption, in turn, would have made it much more difficult in Part II to make sense of the struggles (large and small) around the very possibility of governmental knowledge.

Another advantage of using Foucault's archaeological method was that it allowed me to use a particular individual as a lens on the early discourse of governmentality without investing in all the trappings of the biographical genre. By slowly "assembling" Francis Walker's relevance out of a more structural analysis of subject positions in Chapters 2 and 3, I hoped to make it clear that, however interesting he was as an individual, his career had a *general* relevance to the account of governmentality. I wished to throw up as many barriers as possible against the assumption that this was a study of a "great figure," or that his idiosyncracies were the main point. There were certainly aspects of Walker's personality that set him apart from most other comparable figures of the time, but these, too, have *general* relevance. In Chapter 4, I took somewhat of a detour to explain in depth Walker's unusually strong investment in the late nineteenth-century cult of manliness. But the point of doing so was to be able to highlight the gender-related aspects of his governmental program in Part II, and thereby to illustrate one major gender valence that could attach to such programs. As the literature on women in the Progressive Movement makes plain (see Chapter 3), the connection between manhood and modern regimes of social regulation was by no means automatic, and competing feminist programs had their own very substantial successes. Walker served very well to illustrate the detailed gender logic behind one major "male" approach to modern social control. In short, the "biographical" material I assembled tells us more about Walker's context than about his innermost being, and the archaeological approach proved a suitable means of making this clear.

Contributions to the conceptual framework of governmentality

The particular contributions made by this study to the notion of governmentality have much to do with its focus on census-taking. To put it very briefly, the "grids of specification" most prominent in structuring the US Census (namely, space, race and gender – see Chapter 2) both expressed and constructed the larger cultural "dimensions" in which most late nineteenth-century Americans understood their national society. The census was thus an unusually revealing institution through which to trace the mutual influence of the logic of governmentality and the larger cultural context in which it was embedded. In Part II, I was able not only to reinforce the obvious point that governmentality always acquires its specific form in concrete cultural context, but also to show *how* late nineteenth-century middle-class discourses of race and gender (in particular) interacted with an inherently spatial logic of control to produce Walker's program of immigration restriction.

In the process, it was also possible to suggest a number of different types or levels at which the basic logic of governmentality can operate. For example, an important differentiator is spatial scale. In general, particularly in the American context, *national* programs of governmentality are much less likely than are, say, urban-scale programs, to progress beyond the "program" stage to that of implementation. This is due not only to the vastly greater material and logistical difficulty national states face in controlling the minutiae of social activity throughout their territories, but also to the vastly more cumbersome and uncertain complex of institutions and practices through which any national-scale program must be mediated on the way to implementation. The opportunities for derailment of a national-scale program (especially in a nominally "democratic" society) are often so numerous as to be almost prohibitive in the aggregate. Put differently, national states are still, even today, struggling always to create "the national scale."

Another facet of national-scale governmentality which deserves further consideration is its association (in the American context, at least) with masculinity. The distinction noted in recent feminist scholarship between a "male" brand of early American social science increasingly segregated in academic departments of economics, sociology, psychology and political science, and a "female" brand focused on reform initiatives at the urban scale and centered on settlement houses and other non-academic institutions, is fascinating. Trained to the same levels of expertise as their male counterparts but excluded from full equality in the rising academic disciplines, late nineteenth-century American women social scientists tended to produce research geared much more strongly toward practical reform, and rooted in the particular places where they lived and carried out their work (see Chapters 1 and 3). The programs of regulation characteristic of the Progressive Era are thus seen to have expressed an identifiably "feminine" and feminist tradition of social science. A kind of gendered spatial division of labor emerged within social science that paralleled the ideal division of labor familiar from the Victorian domestic ideology. Men had secured for themselves almost exclusively the national networks of social science departments as well as the production of research concerning national issues. In other words, in keeping with my analysis of Walker's program, men had secured for themselves "national subjectivity." By contrast, women's social science was local and more openly guided by moral concerns. At bottom, this literature leads directly to the conclusion that gender must play an important role not merely in analyses of governmentality but also in the very theorization of it. It would be interesting to compare the American experience in this regard with that of other modern industrial countries.

When the gender issues identified in this study are brought into contact with the analysis of spatial politics, we can see that the basic patriarchal construction of gender dualism serves programs of national territorial mastery by *resolving the contradiction between the need for fixity in the "objects" of control and the need for mobility on the part of the "subjects" of control.* Conceptually domesticated, women could, for Walker, collectively *anchor* the social body at millions of fixed points, not only allowing its generational reproduction but stabilizing it for inspection. Men, though still tied to these domestic points, could command the avenues of circulation necessary for a healthy economy as well as for the restless movement of national authoritative gazes through which official institutions monitor the state of the social body. This sort of logic, too, would prove an interesting angle from which to view various colonial episodes. How have imposed Western gender ideologies interacted with different pre-existing gender regimes in different cultures that have played unwilling host to colonial impositions?

Another interesting set of distinctions within the broad category of governmentality which it would be worthwhile to explore has to do with "layers" of spatial organization. Whether at the national or the urban scale, it is possible to distinguish between "discursive" programs of abstraction, assortment and centralization (that is, programs operating through legal or fiscal means), on the one hand, and the physical infrastructures either produced or used in pursuit of these programs, on the other. And then, within the category of abstraction, there are a number of sublevels, each having a somewhat different role to play in modern governance. All of these levels are more or less interdependent, with some being more fundamental than others. For example, among the preconditions needed to carry out a census there must be at least rudimentary networks of roads and other transportation webs, and in addition, trigonometrical and topographic surveys need to have been done in order to allow unambiguous location. In all probability, land and cadastral surveys must also be in place. Each of these layers, as well as the different layers of infrastructure in cities (sewers, subways, as well as roads), can be thought of as an expression of governmental logic. Each could be understood to a greater or lesser extent through the categories I have used to describe the "spatial politics" of governmentality. However, there are undoubtedly dynamics specific to each layer at each scale which would reward further investigation.

These considerations in turn raise the question of the "fostering" versus "constraining" aspects or types of governmentality. Particularly when scholars use a governmentality perspective in order to understand the building of urban or national infrastructure, it is not so clear that the cycle of social control as I have laid it out is the most appropriate conceptual framework. Governmentality will always involve moments of observation and normalizing judgment, but (as Walker's convoluted arguments regarding "socialism" suggest – see Chapter 7), "regulation" is perhaps not the best word for actions undertaken to enhance social welfare. On the other hand, infrastructural networks always constrain and channel in the process of enabling, and among the activities they enable are activities related to regulation. This is, in any case, an aspect of governmentality that needs to be elaborated further.

Contributions to the historiography of modern American state formation

In order to make a convincing case for the usefulness of a governmentality perspective, I need to be able to point to concrete contributions it can make to current historiography. How does the theory improve interpretation of the empirical material? There are at least three ways. First, such a perspective draws attention to the *constructed* nature of the "national social body" as an object of state concern, as well as to "the state" itself. As I noted in Chapter 1, some Marxist state theorists have begun to arrive at similar conclusions, but from a more exclusively theoretical direction which lends itself less well than a Foucauldian approach to concrete analysis. Once state activities such as census-taking are understood as *constituting* the social body, rather than merely as attempts to learn more about it, it becomes possible to understand a good deal more, for example, about the significance of race and ethnicity in American history and historical geography. "Race" as a set of judicial and social-scientific categories has *made and remade*, not merely "reflected," a crucial aspect of American social order since the eighteenth century.

Even apart from the specifics of such insights, I hope I have made a case for some level of compatibility between Foucauldian and other (for example, Marxist and feminist) analyses of state power. If we can recognize that the constructed character of a state or a national social body does not imply that it lacks considerable stability and effectivity, it becomes possible to understand governmentality as a *complementary* approach. This is not to suggest that all tensions and problems of articulation disappear; it is merely to encourage perseverance in attempting to overcome them. A hybrid framework capable of doing full justice both to the contingency and to the solidity of apparatuses of social control would constitute a significant advance.

A second way in which the theorization of governmentality given in this study improves our interpretive abilities is in its focus on the spatial character of governmental knowledge and regulation. Social theorists such as Henri Lefebvre, Anthony Giddens and Michael Mann have drawn general historical connections between state modernization and territorial mastery; the anthropologist Bernard Cohn, the peasant studies scholar James C. Scott and the geographer Matthew Edney have begun to give analytical flesh to this connection (see Chapter 5). I have attempted here to systematize their insights in a more detailed analysis of the "spatial politics" of territorial control, especially through the discussion in Chapter 5 of "abstraction," "assortment," "centralized control" and "compilation." Chapter 7 shows how the logic of spatial control, operating in an ideological environment which highlighted gender and race as primary dimensions of social order, led Walker logically to a policy of boundary closure.

The spatial analysis offered here is clearly relevant not only to instances of modern state formation in other periods of US history and in other "Western" countries, but also to histories of colonization. Indeed, the fact that such close parallels can be drawn between territorial mastery "at home" and "in the colonies" helps to emphasize and clarify the political (that is, contestable) character of such seemingly benign activities as census-taking. Paul Rabinow's monumental study *French modern: Norms and forms of the social environment* provides an excellent point of comparison here. Rabinow takes up many implicitly and explicitly governmental themes, including spatial politics, in tracing the careers of a set of French officials in the late nineteenth and early twentieth centuries. These officials were instrumental in developing a culture of social and urban planning in the French colonies, and then transferring many of their ideas to social engineering schemes "at home" in a France threatened by social disorders attendant on class struggle. Among many interesting figures, Emile Cheysson (1836–1910) stands out as Francis Walker's French twin: Cheysson served as president of the Société d'Economie Sociale, which sought to establish sociology in the French universities and had a strong presence in the newly formed statistical bureaus of the government. Cheysson, like Walker, had war-related administrative experience, organizing the grain rationing for Paris during the German siege of 1870. He went on to direct an experimental factory, and then for seven years the national Bureau of Maps and Plans, where he was heavily involved in the cadastral mapping of the entire country. From 1887 to 1901, he held the chair of political economy at the Ecole Libre des Sciences Politiques, and was elected president of the Société de Géographie in 1896.[1] According to Rabinow, Cheysson "was one of the leading figures in the organization and propagation of statistical methods in France," serving in 1881 as president of the Société de Statistique de Paris, "where he helped coordinate and consolidate (across different government agencies) existing collections of data and argued forcibly (and successfully) for the inclusion of more industrial statistics."[2]

The parallels with Walker are fascinating, but so are the differences. It might be possible, using such studies comparatively, to isolate the effects on "home" governmentality of the possession of overseas colonial empires, or to compare the role the French gave to the study of colonized peoples in Morocco or Madagascar with the ethnographic projects undertaken by American officials like John Wesley Powell (founder of the Bureau of American Ethnology). Here again, Matthew Edney's study of the British mapping of India would provide an excellent benchmark.[3] In addition, by comparison with European examples, it might be possible to distinguish more clearly the ways in which class failed to become a major "grid of specification" in American social statistics. It might further be possible to link differences in the spatial politics of access to information with the different degrees of centralization characteristic of different (for example, French and American) state systems. Were French citizens more or less willing to submit to censuses and regulations than their American counterparts? How did officials like Cheysson deal with whatever resistance they encountered? Did resistance have a class flavor to it, in contrast to the American case?

The third, and perhaps most immediate historiographical benefit of the present study lies in the challenge it poses to the basic binary distinction at the heart of most studies of modern American state formation during the second half of the nineteenth century. This is the distinction between a liberal, *laissez-faire* ideology (and policy), on the one hand, and on the other a "progressive," regulationist approach. Most interpreters of the political dynamics of this period, including Morton Keller, Sidney Fine and Stephen Skowronek, accept and perpetuate, *even while seeming to deconstruct*, a relatively simple opposition between a basic hostility to regulation in the Gilded Age and the dawn of fairly vigorous "progressive" embrace of regulation around the turn of the century (see Chapter 1). All of these authors acknowledge that significant

---

[1] P. Rabinow, *French modern: Norms and forms of the social environment* (Chicago, IL, 1989), 172.
[2] *Ibid.*, 174.
[3] M. Edney, *Mapping an empire: The geographical construction of British India, 1765–1843* (Chicago, 1997).

regimes of regulation predated the Progressive Era, and that a basic presumption in favor of minimal government survived well after the Gilded Age, but all tend to treat both the early attempts at stronger regulation and the later survivals of *laissez-faire* attitudes as anomalies essentially at odds with the respective spirits of the two periods. They miss the more complicated and subtle logic of governmentality, in part by paying insufficient attention to specifically statistical developments, in part by portraying partisan politics as the determining essence of late nineteenth-century American political reality, and in part by focusing their analyses on "administrative capacity," a concept which tends to obscure the crucial governmental distinction between the ability to regulate and the decision to regulate. Their analyses are not in any sense "wrong"; indeed the studies these authors have produced constitute a rich portrait of this period of American political history. However, as I hope this study has shown, their composite portrait is incomplete.

"Governmentality" as I have developed it here is a name for a logic of social regulation that *consistently blends* the principles of freedom and regulation, rendering the blend itself a distinct and principled approach, and making the question of whether and how much to regulate a matter for case-by-case empirical investigation. "Liberalism" of the twentieth-century variety has come to encompass much of what I have been calling governmentality, but the liberalism of the late nineteenth century was not so subtle. Thus "governmentality" names something not adequately captured by scholars of the period, and allows us to avoid the mild dismissal implicit in the idea that programs such as Walker's were "transitional."

This point was rendered much easier for me to make by the structure of Part II, centered as it was around the "cycle of social control." This cycle is, again, divided into three moments: observation, normalizing judgment and enforcement or regulation. Simply by distinguishing so strongly between observation and regulation, the argument made it far easier to see that the first does not always automatically lead to the second. If we recognize a moment of judgment connecting the two, and recognize in particular the importance of the governmental question *whether* to regulate, it will be much less difficult to break out of the Weberian habit of conflating knowledge and regulation, or power to regulate and regulation, in categories such as "administrative capacity."

It might be objected to all of this that the "liberal" view of regulation already encompasses all the subtlety I attribute to governmentality. Liberalism, one might insist, is founded in a predisposition to *distrust* regulation, but only in the most extreme cases does it become an insistence on the *wholesale abandonment* of regulation. In this view, Francis Walker was nothing more than a relatively sensible liberal. Thus, there is nothing analytically new in the notion of governmentality. To this line of argument I would make the following set of responses. First and most simply, a subtle view of liberalism does not mark the historical interpretations given by the scholars discussed in Chapter 1. However nuanced their understanding of liberalism *in practice*, the stories they tell are fundamentally polarized by the two competing principles of freedom and regulation. The admirable subtlety of these stories comes not in describing a balance between the two principles, but in describing the historical factors working for and against each of them. The two principles remain the chief focus. A governmentality perspective encourages analysis not of the way the two principles *compete* in concrete historical settings, but of the way the principle of freedom is brought into contact with empirical data, and of the way regulatory decisions emerge out of this contact.

What I have been trying to do with these schematic comments is to flesh out the concept of governmentality in ways that will help it perform the offices of what Foucault terms "general history" (see the introduction to Part II). The main point of general history, once again, is to elaborate analytical "spaces of dispersion" in which to compare a wide range of phenomena without unduly reducing the specificity of each instance or case. As a more sophisticated theorization of governmentality emerges from an ever-growing roster of individual studies, we should not lose sight of the need to ensure that this theorization remains analytically useful.

Bibliography

United States Government Documents

Department of Commerce, Bureau of the Census, *Historical statistics of the United States*, 2 vols., Washington, DC: GPO, 1975.

Government Printing Office, *Annual Report of the Public Printer*, Senate Misc. Documents 12, 47th C., 2nd S., US Serial Set 2083, Washington DC: GPO 1882.

Government Printing Office, *Annual Report of the Public Printer*, Senate Misc. Documents 10, 48th C., 1st S., US Serial Set 2170, Washington DC: GPO, 1883.

"Letter from the Secretary of the Interior submitting a report of the receipt and distribution of public documents on behalf of the Government by the Department of the Interior," Feb. 17, 1886, House Executive Documents 78, 49th C., 1st S., US Serial Set 2398, Washington DC: GPO, 1886.

"Letter from the Secretary of the Interior transmitting a report by the Superintendent of Documents regarding receipt, distribution and sale of public documents on behalf of the Government," Feb. 18, 1890, House Executive Documents 212, 51st C., 1st S., 1890.

National Archives and Records Administration (NARA), *Preliminary inventories No. 161: Records of the Bureau of the Census*, Davidson, K. and Ashby, C. (compilers), Washington, DC: Government Printing Office, 1964.

Record Group 29, Records of the Bureau of the Census, Administrative Records of the Census Office.

Office of the Census, *Annual Report of the Superintendent of the Census*, House Executive Documents 1, 47th C., 1st S., 1881, 665–727.

Ninth Census: The statistics of the population of the United States, vol. I, Washington, DC: GPO, 1872.

Statistics of the population of the United States at the Tenth Census, Walker, F. and Seaton, C. (compilers), Washington, DC: GPO, 1883.

US Congress, "An Act making appropriations for sundry civil expenses of the Government," July 7, 1884, *US Statutes at Large*, vol. 23, 48th C., 194–226.

"An act to provide for the publication of the Tenth Census," Aug. 7, 1882, *US Statutes at Large*, vol. 22, 47th C., 1882, 344–345.

"An act providing for the taking of the Seventh and subsequent Censuses," May 23, 1850, *US Statutes at Large*, vol. 9 (31–1[1850]), ch. 11, 428–436.

"An act to provide for taking the Tenth and subsequent Censuses," Mar. 3, 1879, *US Statutes at Large,* vol. 20, 45th C., 3rd S., 1879, Ch. 195, 473–481.

Report of the House Committee on Printing, House Reports 1784, 47th. C., 1st S., US Serial Set 2151, Washington, DC: GPO, 1882.

US Immigration Commission, *Abstracts of reports of the Immigration Commission,* Senate Document 747, 61st C., 3rd S, US Serial Set 5865, Washington, DC: GPO, 1911.

"Statement of the Immigration Restriction League," in *Statements and recommendations submitted by societies and organizations interested in the subject of immigration,* Senate Document 764, 61st C., 3rd S., US Serial Set 5881, Washington, DC: GPO, 1911.

Walker, F., "Instructions to assistant marshals," May 1, 1870, reprinted in Wright, C. and Hunt, W., *History and growth of the United States Census,* 1900, 155–166.

"Instructions to enumerators" May 1, 1880, reprinted in Wright, C. and Hunt, W., *History and growth of the United States Census,* 1900, 167–177.

"Preface and introduction to the statistical atlas," in *Statistical atlas of the United States,* Washington, DC, 1874, n.p.n.

"The progress of the nation, 1790–1870," in *Statistical atlas of the United States,* Washington, DC, 1874, n.p.n.

"The relations of race and nationality to mortality in the United States," in *Statistical atlas of the United States,* Washington, DC, 1874, n.p.n.

"Report of the Superintendent of the Census to the Secretary of the Interior," Dec. 26, 1871, in Office of the Census, *Ninth Census,* vol. 1: *The statistics of the population of the United States,* 1872, n.p.n.

"Report of the Superintendent of the Ninth Census," Aug. 24, 1872, in Office of the Census, *Ninth Census,* vol. 1: *The statistics of the population of the United States,* 1872, n.p.n.

Walker, F. (comp.), *Statistical atlas of the United States,* Washington, DC: Julius Bien, 1874.

Wright, C. D., and Hunt, W., *History and growth of the United States Census,* Senate Document 194, 56th C., 1st S., US Serial Set 3856, Washington, DC: GPO, 1900.

Francis A. Walker sources (exclusive of government documents)

Francis A. Walker Papers, Library of Congress, Manuscript Reading Room.

Francis A. Walker Papers, 1862–1897, Massachusetts Institute of Technology Archives and Special Collections, MC 298

President's reports, MIT Archives and Special Collections, public reading room.

Walker, F., "Address to the International Statistical Institute," from *Bulletin de L'Institue International de Statistique* 8 (1895), xxxvi–xxxix, reprinted in *DES,* vol. II, 141–145.

"American agriculture," *Princeton Review* 9 (1882), 249–264, reprinted in *DES,* vol. II, 157–175.

"American industry in the Census," *Atlantic Monthly* 24 (1869), 689–701, reprinted in *DES,* vol. II, 3–25.

"American manufactures," *Princeton Review* 11 (1883), 213–223, reprinted in *DES,* vol. II, 179–190.

"The bases of taxation," *Political Science Quarterly* 3 (1888), 1–16, reprinted in *DES,* vol. I, 79–94.

"Cairnes' 'Political economy'," review of J. E. Cairnes, *Leading principles of political economy*, reprinted in *DES,* vol. I, 279–280.

"The causes of poverty," *The Century* 55 (1897), 210–216, reprinted in *DES,* vol. II, 455–471.

"Census," in *Encyclopaedia Britannica*, 9th ed., vol. V, New York: Charles Scribner's Sons, 1876, 334–340.

"The colored race in the United States," *The Forum* 11 (1891), 501–509, reprinted in *DES,* vol. II, 127–137.

"Defects of the census," from "Report of the Superintendent of the Census," Nov. 1, 1874, *House Executive Documents* 1, part 5 (43rd C., 2nd S.), 724–729, reprinted in *DES,* vol. II, 49–58.

Discussions in Economics and Statistics [*DES*], 2 vols., Dewey, D. (ed.), New York: Henry Holt and Co., 1899.

Discussions in Education [*DE*], Dewey, D. (ed.), New York: Henry Holt and Co., 1898.

"The eight-hour law agitation," a lecture delivered before the Boston Young Men's Christian Union, Mar. 1, 1890, reprinted in *DES,* vol. II, 379–396.

"The Eleventh Census of the United States," *Quarterly Journal of Economics* 2 (1888), 135–161, reprinted in *DES,* vol. II, 79–80.

"Enumeration of the population, 1870–1880," *Publications of the American Statistical Association* 2 (1890), n.p.n, reprinted in Walker, *DES,* vol. II, 61–65.

General Hancock, New York: D. Appleton and Co., 1895.

"The great count of 1890," *The Forum* 11 (1891), 406–418, reprinted in *DES,* vol. II, 111–124.

"The growth of American nationality," *The Forum* 20 (1895), in Francis A. Walker Papers, 1862–97, Massachusetts Institute of Technology Archives and Special Collections, MC 298.

"Growth of the nation," from Phi Beta Kappa oration, Brown University, June 18, 1889, reprinted in *DES,* vol. II, 193–212.

"Immigration and degradation," *The Forum* 11 (1891), 634–643, reprinted in *DES,* vol. II, 417–426.

"The Indian question," *North American Review* 66, 239 (1873), 329–388.

"Industrial education," address before the American Social Science Association, Sept. 6, 1884, reprinted in *DE,* 125–149.

"Interview of the Select Committees of the Senate of the United States and of the House of Representatives to make provisions for taking the Tenth Census," Dec. 27, 1878, *Senate Misc. Documents* 26 (45th Congress, 3rd Session), in Francis A. Walker Papers, Massachusetts Institute of Technology Archives and Special Collections, MC 298, Box 3, Folder 28.

"The laborer and his employer," lecture delivered at Cornell University, Feb. 1889, *Scientific American*, June 1, 1889, supplement no. 700, typescript, Francis A. Walker Manuscript Collection, MIT Archives and Special Collections MC 298, Box 4, Folder 3.

The making of the nation, 1783–1817, New York: Charles Scribner's Sons, 1897.

"Manual education in urban communities," address before the National

Educational Association, Chicago, IL, July 15, 1887, reprinted in Walker, F., *DE*, 175–194.

"Methods of restricting immigration," *Yale Review* 1 (1892), 138–143, reprinted in *DES*, vol. II, 429–434.

Money, Baltimore, MD: Johns Hopkins University Press, 1878.

"Normal training in women's colleges," *Educational Review* (1892), reprinted in *DE*, 305–319.

"Occupations and mortality of our foreign population, 1870," from *Chicago Advance* Nov. 12, 1874, Dec. 10, 1874, Jan. 14, 1875, reprinted in *DES*, vol. II, 215–222.

"Our domestic service," *Scribner's Monthly* 11 (1875), 273–278, reprinted in *DES*, vol. II, 225–237.

"Our population in 1900," *Atlantic Monthly* 32 (1873), 487–495, reprinted in *DES*, vol. II, 29–45.

"A plea for industrial education in the public schools," address before the Conference of Associated Charities of the City of Boston, Feb. 10, 1887, reprinted in *DE*, 153–172.

Political economy, 3rd ed., New York: Henry Holt and Co., 1888.

"Political geography and statistics," in *Encyclopaedia Britannica*, 9th ed., vol. xxiii, New York: Charles Scribner's Sons, 1888, 818–829.

"The present standing of political economy," *Sunday Afternoon* 3 (1879), 432–441, reprinted in *DES*, vol. I, 301–318.

"Private property," unpublished lecture, n.d., reprinted in *DES*, vol. II, 405–411.

"Protection and protectionists," *Quarterly Journal of Economics* 4 (1890), 245–275, reprinted in *DES*, vol. I, 97–124.

"The relation of professional and technical to general education," *Educational Review* (1894), reprinted in *DE*, 55–78.

"Restriction of immigration," *Atlantic Monthly* 77 (1896), 822–829, reprinted in *DES*, vol. II, 437–451.

"The rise and importance of applied science in American education," address at the Convocation of the University of the State of New York, Albany, NY, July 9, 1891, reprinted in *DE*, 19–35.

"Socialism," *Scribner's Magazine* 1 (1887), 107–119, reprinted in *DES*, vol. II, 247–271.

"The source of business profits," *Quarterly Journal of Economics* 1 (1887), 265–288, reprinted in *DES*, vol. I, 359–380.

"The United States Census," *The Forum* 11 (1891), 258–267, reprinted in *DES*, vol. II, 97–107.

Wages: A treatise on wages and the wages class, New York: Henry Holt and Company, 1876.

"What shall we tell the working classes?," *Scribner's Magazine* 2 (1887), 619–627, reprinted in *DES*, vol. II, 301–317.

Works cited (general)

Anderson, B. *Imagined communities*, revd. edn., New York: Verso, 1991.

Anderson, M. *The American census: A social history,* New Haven, CT: Yale University Press, 1988.

(Anonymous), "A plea for the protection of useful men," *Review of Reviews*, Feb. 1897 (n.p.), in Francis A. Walker Papers, Massachusetts Institute of Technology Archives and Special Collections, MC 298, Box 5.

Aron, C. S., *Ladies and gentlemen of the civil service: Middle-class workers in Victorian America*, New York: Oxford University Press, 1987.

Athearn, R., *Union Pacific country*, Lincoln, NE: University of Nebraska Press, 1976.

Barry, A., Osborne, T. and Rose, N. (eds.), *Foucault and political reason: Liberalism, neo-liberalism and rationalities of government*, Chicago: University of Chicago Press, 1996.

Barry, A., Osborne, T. and Rose, N., "Introduction," in Barry, A., Osborne, T. and Rose, N. (eds.) *Foucault and political reason*, 1996, 1–18.

Bensel, R., *Yankee leviathan: The origins of central state authority in America, 1859–1877*, New York: Cambridge University Press, 1990.

Blaug, M., *Economic history and the history of economics*, New York: New York University Press, 1986.

Blaut, J., *The colonizers' model of the world: Geographical diffusionism and Eurocentric history*, New York: Guilford Press, 1993.

Brubacher, J. and Rudy, W., *Higher education in transition: A history of American colleges and universities*, 4th ed., New Brunswick, NJ: Transaction Publishers, 1996.

Burchell, G. "Peculiar interests: Civil society and governing 'the system of natural liberty'," in Burchell *et al.* (eds.), *The Foucault effect*, 1991, 119–150.

Burchell, G., Gordon, C. and Miller, P. (eds.), *The Foucault Effect: Studies in governmentality*, Chicago: University of Chicago Press, 1991.

Chielens, E. (ed.), *American literary magazines*, Westport, CT: Greenwood Press, 1986.

Cohen, P.C., *A calculating people: The spread of numeracy in early America*, Chicago: University of Chicago Press, 1982.

Cohn, B. "The census, social structure and objectification in South Asia," in *An anthropologist among the historians and other essays*, New York: Oxford University Press, 1987, 224–254.

Cohn, B., *Colonialism and its forms of knowledge: The British in India*, Princeton, NJ: Princeton University Press, 1996.

Conk, M. "The Census, political power, and social change: The significance of population growth in American history," *Social Science History* 8 (1984), 81–106.

Coontz, S., *The social origins of private life: A history of American families, 1600–1900*, New York: Verso, 1988.

Corner, J. and MacLean, A., *Taking measures across the American landscape*, New Haven, CT: Yale University Press, 1996.

Cremin, L., *American education*, vol. III: *The metropolitan experience, 1876–1980*, New York: Harper and Row, 1988.

Cummings, J., "Statistical work of the Federal Government of the United States," in Koren, J. (ed.), *The history of statistics: Their development and progress in many countries*, New York: American Statistical Association, 1918, 571–689.

Daniels, G., (ed.), *Nineteenth century American science: A reappraisal*, Evanston, IL: Northwestern University Press, 1972.

Davis, R. C., "The beginnings of American social research," in Daniels, G. (ed.), *Nineteenth century American science*, 1972, 152–178.

Deegan, M. J., *Jane Addams and the men of the Chicago School, 1892–1918*, New Brunswick, NJ: Transaction Publishers, 1988.

Deleuze, G., *Foucault*, Minneapolis, MN: University of Minnesota Press, 1988.

Dirks, N., "Foreword," in Cohn, B., *Colonialism and its forms of knowledge*, 1996, ix–xvii.

Donzelot, J., *The policing of families*, Baltimore, MD: Johns Hopkins University Press, 1997 [1979].

Dreyfus, H. and Rabinow, P., *Michel Foucault: Beyond structuralism and hermeneutics*, 2nd ed., Chicago: University of Chicago Press, 1983.

Driver, F., "Moral geographies: Social science and the urban environment in mid-nineteenth century England," *Transactions of the Institute of British Geographers* 13 (1987): 275–287.

"Political geography and state formation: disputed territory," *Progress in Human Geography* 15, 3 (1991), 268–280.

"Bodies in Space: Foucault's account of disciplinary power," in Jones, C. and Porter, R. (eds.), *Re-Assessing Foucault*, New York: Routledge, 1992.

Power and pauperism: The workhouse system, 1834–1880, New York: Cambridge University Press, 1993.

Dupree, A.H., "The measuring behavior of Americans," in Daniels, G. (ed.), *Nineteenth century American science*, 1972, 22–37.

Edney, M., *Mapping an empire: The geographical construction of British India, 1765–1843*, Chicago: University of Chicago Press, 1997.

Ekelund, R. B., "Contributions of Francis Amasa Walker to economic thought," MA thesis, San Antonio, TX: St. Mary's University, 1963.

Elkins, S. and McKittrick, E., *The age of federalism*, New York: Oxford University Press, 1993.

Fine, S., *Laissez-faire and the general welfare state: A study of conflict in American thought, 1865–1901*, Ann Arbor, MI: University of Michigan Press, 1956.

Fitzpatrick, E., *Endless crusade: Women social scientists and progressive reform*, New York: Oxford University Press, 1990.

Fletcher, W. I., "Preface," in Poole, W F. and Fletcher, W. I., (comps.), *Poole's index to periodical literature: The second supplement*, Boston, New York: Houghton, Mifflin and Co., 1893, iii–v.

"Obituary for William F. Poole," in Poole, W. F. and Fletcher, W. I., (comps.), *Poole's index to periodical literature: The third supplement*, 1897, iii–iv.

Foucault, M., *The archaeology of knowledge*, New York: Pantheon, 1972.

Discipline and punish, New York: Vintage, 1977.

The history of sexuality, vol. I, An introduction, New York: Vintage, 1978.

"On governmentality," *Ideology and consciousness* 6 (1979), 5–21.

"Truth and power," in Gordon, C. (ed.), *Power/Knowledge: Selected interviews and other writings, 1972–1977*, New York: Pantheon, 1980, 109–133.

The use of pleasure: The history of sexuality, vol. II, New York: Vintage, 1986.

The care of the self: The history of sexuality, vol. III, New York: Vintage, 1988.

"Governmentality," in Burchell *et al.* (eds.), *The Foucault effect*, 1991, 87–104.

Fredrickson, G. M., *The inner Civil War: Northern intellectuals and the crisis of the Union*, New York: Harper and Row, 1965.

Giddens, A., *The nation-state and violence*, Berkeley, CA: University of California Press, 1987.

Daniel Coit Gilman Papers, Special collections, Johns Hopkins University, Baltimore, MD, MS 001.

Gordon, C., "Governmental rationality: An introduction," in Burchell, G., Gordon, C. and Miller, P. (eds.), *The Foucault Effect*, 1991, 1–52.

Gossett, T., *Race: The history of an idea in America*, New York: Schocken Books, 1989 [1963].

Gregory, D., *Geographical imaginations*, New York: Blackwell, 1994.

Hannah, M. "Space and social control in the administration of the Oglala Lakota ("Sioux"), 1871–1879," *Journal of Historical Geography* 19, 4 (1993); 412–432.

"Imperfect panopticism: Envisioning the construction of normal lives," in Benko, G. and Strohmayer, U. (eds.), *Space and social theory: Interpreting modernity and postmodernity*, New York: Blackwell, 1997, 344–359.

"Space and the structuring of disciplinary power: An interpretive review," *Geografiska Annaler* 79 B, 3 (1997), 171–180.

Harley, J. B., "Deconstructing the map," *Cartographica* 26 (1989), 1–20.

Haury, D., "*Niles Weekly Register,*" in Nourie, A. and Nourie, B. (eds.), *American mass-market magazines*, 1990, 330–332.

Heilbroner, R., *The worldly philosophers: The lives, times and ideas of the great economic thinkers*, 6th ed., New York: Touchstone, 1986.

Higham, J., "The matrix of specialization," in A. Oleson and J. Voss, *Organization of knowledge*, 1979, 3–18.

Strangers in the land: Patterns of American nativism, 1860–1925, New Brunswick, NJ: Rutgers University Press, 1994 [1955].

Hilkey, J., *Character is capital: Success manuals and manhood in Gilded Age America*, Chapel Hill, NC: University of North Carolina Press, 1997.

Hofstadter, R., *Anti-intellectualism in American life*, New York: Vintage, 1962.

Social Darwinism in American thought, Boston: Beacon Press, 1992 [1944].

Hunt, C., *The life of Ellen H. Richards*, Boston: Whitcomb and Barrows, 1912.

Jessop, B., *State theory: Putting capitalist states in their place*, University Park, PA: Penn State Press, 1990.

Keller, M., *Affairs of state: Public life in late nineteenth century America*, Cambridge, MA: Harvard University Press, 1977.

Regulating a new economy: Public policy and economic change, 1900–1933, Cambridge, MA: Harvard University Press, 1990.

Kendrick, J. W., "The economics of Francis Walker," MA thesis, Chapel Hill, NC: University of North Carolina, 1939.

Kerr, R. W., *History of the Government Printing Office (at Washington DC), with a brief record of the public printing for a century, 1789–1881*, New York: Burt Franklin, 1970 [1882].

Kimmel, M., *Manhood in America: A cultural history*, New York: Free Press, 1996.

Kohut, D., "*DeBow's Review,*" in Nourie, A. and Nourie, B. (eds.), *American mass-market magazines*, 1990, 95–97.

Koren, J., "The American Statistical Association," in Koren, J. (ed.), *The history of statistics: Their development and progress in many countries*, New York: American Statistical Association, 1918, 3–14.

Koven, S. and Michel, S. (eds.), *Mothers of a new world: Maternalist politics and the origins of welfare states*, New York: Routledge, 1993.

Latour, B., *Science in action*, Cambridge, MA: Harvard University Press, 1987.

We have never been modern, Cambridge, MA: Harvard University Press, 1993.

Lears, T. J. Jackson, *No place of grace: Antimodernism and the transformation of American culture, 1880–1920*, Chicago: University of Chicago Press, 1983.

Lee, A. M., *The daily newspaper in America: The evolution of a social instrument*, New York: Farrar, Strauss, Giroux, 1973.

Littlefield, D. and Underhill, L., "Renaming the American Indian: 1890–1913," *American Studies* 12, 2 (1971), 33–45.

Livermore, T., "In memoriam – Companion Brevet Brigadier General Francis Amasa Walker," Military Order of the Loyal Legion of the United States, Commandery of the State of Massachusetts, Circular no. 2, series 1897, in Francis A. Walker Papers, MIT Archives and Special Collections, MC 298, Box 5.

Livingstone, D., *Nathaniel Southgate Shaler and the culture of American science*, Tuscaloosa, AL: University of Alabama Press, 1987.

Lucas, C., *American higher education: A history*, New York: St. Martin's Press, 1994.

Lutz, T., *American nervousness, 1903*, Ithaca, NY: Cornell University Press, 1991.

Mangan, J.A. and Walvin, J. (eds.), *Manliness and morality: Middle-class masculinity in Britain and America, 1800–1940*, New York: St. Martin's Press, 1987.

Mann, M., *The sources of social power*, vol. I, *A history of power from the beginning to A.D. 1760*, New York: Cambridge University Press, 1986.

  The sources of social power, vol. II, *The rise of classes and nation-states, 1760–1914*, New York: Cambridge University Press, 1993.

McLaren, A., *The trials of masculinity: Policing sexual boundaries, 1870–1930*, Chicago: University of Chicago Press, 1997.

McNulty, P., *The origins and development of labor economics*, Cambridge, MA: MIT Press, 1980.

Meyer, A., *An educational history of the American people*, New York: McGraw-Hill, 1957.

Meyer, D., "The national integration of regional economies, 1860–1920," in Mitchell, R. and Groves, P. (eds.), *North America: The historical geography of a changing continent*, Totowa, NJ: Rowman and Littlefield, 1987, 321–346.

Montgomery, D., *The fall of the house of labor*, New York: Cambridge University Press, 1987.

Mott, F. L., *A history of American magazines,* 5 vols., New York: D. Appleton and Co., 1930 [vol.I]; Cambridge, MA: Harvard University Press, 1938 [vols. II, III]; 1957 [vols. IV, V].

Mrozek, D. J., "The habit of victory: The American military and the cult of manliness," in Mangan, J. A. and Walvin, J., *Manliness and morality*, 1987, 220–241.

Munroe, J., *A life of Francis Amasa Walker*, New York: Henry Holt and Co., 1923.

Murdoch, J. and Ward, N., "Governmentality and territoriality: The statistical manufacture of Britain's 'national farm'," *Political Geography* 16, 4 (1997), 307–324.

Nelson, D., *National manhood: Capitalist citizenship and the imagined fraternity of white men*, Durham, NC: Duke University Press, 1998.

New York Times Index: A book of record, 1880–1885, New York Times, 1966.

Newton, B., *The economics of Francis Amasa Walker: American economics in transition*, New York: Augustus M. Kelley, 1968.

North, S. N. D., "Seventy five years of progress in statistics: The outlook for the future," in Koren, J. (ed.), *The history of statistics,* 1918, 15–49.

Nourie, A. and Nourie, B. (eds.), *American mass-market magazines*, Westport, CT: Greenwood Press, 1990.

Oleson, A. and Voss, J. (eds.), *The organization of knowledge in modern America, 1860–1920,* Baltimore, MD: Johns Hopkins University Press, 1979.

"Introduction," in Oleson, A. and Voss, J. (eds.), *The organization of knowledge,* 1979, vii–xxi.

Osborne, T., "Security and vitality: Drains, liberalism and power in the nineteenth century," in Barry, A., *et al.* (eds.), *Foucault and political reason,* 1996, 99–122.

Park, R. J., "Biological thought, athletics and the formation of a 'man of character': 1830–1900," in Mangan, J. A. and Walvin, J. (eds.), *Manliness and morality,* 1987, 7–34.

Philo, C., "Foucault's Geography," *Environment and Planning D: Society and Space* 10 (1992), 137–161.

Poole, W. F., *Poole's index to periodical literature,* 3rd ed., Boston: James R. Osgood, 1882.

"Preface," in *Poole's index to periodical literature,* 3rd ed., Boston, 1882, iii–xix.

Poole, W. F. and Fletcher, W. I., "Preface," in *Poole's index to periodical literature: The first supplement,* Boston, New York: Houghton, Mifflin and Co., 1888, iii–iv.

Poole, W. F. and Fletcher, W. I., (comps.), *Poole's index to periodical literature: The third supplement,* Boston, New York: Houghton, Mifflin and Co., 1897.

Poovey, M., *A history of the modern fact: Problems of knowledge in the sciences of wealth and society,* Chicago: University of Chicago Press, 1998.

Porter, T., *The rise of statistical thinking, 1820–1900,* Princeton, NJ: Princeton University Press, 1986.

Trust in numbers: The pursuit of objectivity in science and public life, Princeton, NJ: Princeton University Press, 1995.

Prevost, J.-G., "Francis Walker's theory of immigration and the birth rate: An early twentieth-century controversy," Cahier no. 9701, Cahiers d'épistémologie, Groupe de Recherche en Épistémologie, Montreal, Quebec: Université du Québec à Montréal, 1997.

Procacci, G., "Social economy and the government of poverty," in Burchell, G., *et al., The Foucault effect,* 1991, 151–168.

Prucha, F., *The great father,* 2 vols., Lincoln, NE: University of Nebraska Press, 1984.

Rabinow, P., *French modern: Norms and forms of the social environment,* Chicago: University of Chicago Press, 1989.

Riis, J., *How the other half lives,* New York: Hill and Wang, 1957 [1890].

Roll, E., *A history of economic thought,* 5th ed., Boston: Faber and Faber, 1992.

Rose, G., *Feminism and geography: The limits of geographical knowledge,* Minneapolis, MN: University of Minnesota Press, 1993.

Rose, N. and Miller, P., "Political power beyond the state: Problematics of government," *British Journal of Sociology* 43, 2 (1992), 173–205.

Ross, D., "The development of the social sciences," in Oleson, A. and Voss, J. (eds.), *The organization of knowledge,* 1979, 107–138.

The origins of American social science, New York: Cambridge University Press, 1991.

Rotundo, E.A., *American manhood: Transformations in masculinity from the revolution to the modern era,* New York: Basic Books, 1993.

Rudolph, F., *The American college and university: A history,* New York: Alfred A. Knopf, 1968.

Said, E., *Orientalism,* New York: Vintage, 1979.

Saxton, A., *The rise and fall of the white republic,* New York: Verso, 1990.

Schmeckebier, L., *The Government Printing Office: Its history, activities and organization*, Baltimore, MD: Johns Hopkins University Press, 1925.

Schumpeter, J., *History of economic analysis*, New York: Oxford University Press, 1954.

Schwartz, V., "*Review of Reviews*," in Nourie, A. and Nourie, B. (eds.), *American mass market magazines*, 1990, 435–438.

Scott, J. C., *Seeing like a state: How certain schemes to improve the human condition have failed*, New Haven, CT: Yale University Press, 1998.

Scott, J. W., *Gender and the politics of history*, New York: Columbia University Press, 1985.

Sedgwick, E., "The Atlantic Monthly," in Chielens, E. (ed.), *American literary magazines*, Westport, CT, 1986, 51–54.

Shi, D., *Facing facts: Realism in American thought and culture, 1850–1920*, New York: Oxford University Press, 1995.

Shils, E., "The order of learning in the United States: The ascendancy of the university," in Oleson, A. and Voss, J. (eds.), *The organization of knowledge*, 1979, 19–47.

Sibley, D., *Geographies of exclusion: Society and difference in the West*, New York: Routledge, 1995.

Sidwell, R., "The economic doctrines of Francis Amasa Walker: An interpretation," Ph.D. thesis, University of Utah, 1972.

Silberman, B., *Cages of reason: The rise of the rational state in France, Japan, the United States and Great Britain*, Chicago: University of Chicago Press, 1993.

Silverberg, H., "Introduction: Toward a gendered social science history," in Silverberg, H. (ed.), *Gender and American social science*, 1998, 3–32.

Silverberg, H. (ed.), *Gender and American social science*, Princeton, NJ: Princeton University Press, 1998.

Sklar, K. Kish, "The historical foundations of women's power in the creation of the American welfare state, 1830–1930," in Koven, S. and Michel, S. (eds.), *Mothers of a new world*, 1993, 43–93.

"*Hull House Maps and Papers*: Social science as women's work in the 1890s," in Silverberg, H. (ed.), *Gender and American social science*, 1998.

Skocpol, T., *Protecting mothers and soldiers: The political origins of social policy in the United States*, Cambridge, MA: Harvard University Press, 1992.

Skowronek, S., *Building a new American state: The expansion of national administrative capacities, 1877–1920*, New York: Cambridge University Press, 1982.

Slotkin, R., *The fatal environment: The myth of the frontier in the age of industrialization, 1800–1890*, Middletown, CT: Wesleyan Press, 1985.

Solomon, B., *Ancestors and immigrants*, Cambridge, MA: Harvard University Press, 1956.

Stage, S., "Ellen Richards and the social significance of the Home Economics movement," in Stage, S. and Vincent, V. (eds.), *Rethinking home economics: Women and the history of a profession*, Ithaca, NY: Cornell University Press, 1997, 17–33.

Stearns, P., "Men, boys and anger in American society, 1860–1940," in Mangan, J. A. and Walvin, J. (eds.), *Manliness and morality*, 1987, 75–91.

Straubel, D., "*North American Review*," in Nourie and Nourie *American mass market magazines*, 1990,

Takaki, R., *A different mirror: A history of multicultural America*, Boston: South End Press, 1993.

Townshend, K., *Manhood at Harvard: William James and others,* New York: W. W. Norton, 1996.

Turner, F. J., "The significance of the frontier in American history," in Billington, R. A. (ed.), *Frontier and section: Selected essays of Frederick Jackson Turner,* Englewood Cliffs, NJ: Prentice-Hall, 1961, 37–62.

Walton, G. and Robertson, R., *History of the American economy,* 5th ed., New York: Harcourt Brace Jovanovich, 1983.

Ward, D., *Cities and immigrants: A geography of change in nineteenth century America,* New York: Oxford University Press, 1971.

"Population growth, migration, and urbanization, 1860–1920," in Mitchell, R. and Groves, P. (eds.), *North America: The historical geography of a changing continent,* Totowa, NJ: Rowman and Littlefield, 1986, 299–320.

White, C., *A history of the rectangular survey system,* Washington, DC: Government Printing Office, 1983.

White, L., *The Republican era, 1869–1901: A study in administrative history,* New York: Macmillan, 1958.

Wiebe, R., *The search for order, 1877–1920,* New York: Hill and Wang, 1967.

Winichakul, T., *Siam mapped: A history of the geo-body of a nation,* Honolulu, HI: University of Hawaii Press, 1994.

Wishart, D., "The selectivity of historical representation," *Journal of Historical Geography* 23, 2 (1997), 111–118.

Wright, C. D., "Francis Amasa Walker," *Publications of the American Statistical Association,* new series 38, 1897, 245–275.

Index

abstraction, 39, 107, 113, 124, 142, and US census, 117–123, contradiction with assortment, 127–130
Adams, C. F., 95, 97, 107
Addams, J., 74
administrative capacity, 36, 37, 227
advanced degrees, growth in numbers of American, 71
African Americans, in *Statistical Atlas* of 1874, 152; in Walker's racial thinking, 177, 178, 209
Agassiz, L., 69
American Association for the Advancement of Science, 81
American Economic Association, 13, 70, 74, 81
American exceptionalism, 9, 148, 176
American Geographical Society, 81
American Historical Association, 70, 81
American Indians, in Walker's administrative experience, 208–213; in Walker's racial thinking, 177, 210–212
American Social Science Association, 53, 69, 70, 76, 81
American Statistical Association, 13, 54–55, 69, 81, 217
Amherst College, 78
Anderson, B., 116
Anderson, M., 57, 217
archaeology, and structure of the argument, 11–12, 41, 42, 59, 60, 220–221; and historiography, 41–42; critiques of, 41–42
Arena, 65, 66, 68
Aron, C. S., 140
Arthur, C. A., 52–53
Asian-Americans, in Walker's racial thinking, 178, 209
assortment, 39, 107, 113, 117–118, 124, 142; and US census, 124–130; contradiction with abstraction, 127–130
Atlantic Monthly, 53, 65, 66, 80

authorities of delimitation, 12, 43, 49–56, 61

Barry, A., 76
Bellamy, E., 163
Bensel, R., 35
Bentham, J., 18
best men, 62, 68, 69; and capitalism, 73; detachment of, 74; geographical distribution of, 64, 69; monthly magazines as vehicles for, 66; typical characteristics of, 53; Walker as one of, 78, 81, 99, 107
Blaug, M., 160
Board of Indian Commissioners, 69
borders, in exclusionary programs, 108, 214–215, 219
Boston, as center of magazine publishing, 65, 67
Brahmins (Boston), 78, 215
Breckenridge, S., 74
Bureau of Indian Affairs, and corruption, 69, 79–80, 134, 209, 210

Cairnes, J. E., 162, 163
capitalism, 83; and early American social science, 73; in Foucault's thought, 21; in Walker's thought, 162–163
cartography, critical analysis of, 13; role in administrative control of territory, 114–117,
census (US), 35, 36; as involving quasi-military organizational tasks, 108; as strain on governmental manhood, 101–102; as technology of observation, 13, 39, 117–141; competency exams for employees, 137–138; distribution of results in late nineteenth-century America, 44–49; epistemological structure of, 56–59; growth in size of, 58; hiring of employees, 136–137; methods of enumeration, 126–133; permanent office for, 36, 134, 135;

resistance to, 121–123; role of women in central office, 102, 139–140; scandal of 1840, 51–52; surviving administrative records from late nineteenth century, 136
censuses, political nature of, 114–117
centralized control, 39, 130–134
Century, 53, 65, 80
Cheysson, E., 225, 226
Chicago, University of, 71, 74, 75
Chinese Exclusion Act, 92
cities, invisibility in *Statistical Atlas* of 1874, 144–145; as anchors of governmental discursive formation, 67–71
Civil Service Commission, 54, 69
Civil Service reform, 35, 36, 37, 51–54, 69–70
civil society, 24
Civil War (American), and American population growth rate, 183; and culture of American manhood, 97–98, 198; and state expansion, 34; as incubator of Civil Service reform movement, 53, 61–62; as foundation of American unity, 192
Clark, J. B., 173
Clark University, 71, 94
class, differences in manhood ideals, 89; in late nineteenth-century social order, 85; invisibility in American official statistics, 226; invisibility in *Statistical Atlas* of 1874, 144–145, 148; in Walker's political economy, 175–176, 181, 213
Cohen, P. C., 51
Cohn, B., 116, 117, 225
college graduates, numbers of in late nineteenth century, 62
colonialism, role of knowledge in, 113, 114–117; and regulation, 213, 223, 225–226
compilation, 107, 135–141; as embodied in *Statistical Atlas* of 1874, 141–149
Comtean technocracy, 197
Conk, M., 216
connaissance, 26–27, 43
Coontz, S., 76
Cremin, L., 202
Curtis, G. W., 53
cycle of social control, formal definition, 10–11, 18, 227; role in narrative, 11, 13, 25, 60, 111, 186, 188–189; and discursive formation of American governmentality, 77, 82, 186

de Bow, J. D. B., 47, 55
De Bow's Review, 47, 55
Deegan, M. J., 75
Della Vos, V., 202
Dewey, D. R., 203
discursive formation, 11–12, 40, 42; of early American governmentality, 76–77

Donzelot, J., 75
Driver, F., on pauperism, 8–9,
Dunbar, J. C., 71

Edney, M., 9, 13, 117, 120, 130, 135, 142, 225, 226
education, for women, 200–201, 205–207; industrial, 197–201; traditional elite, 62–63; Walker's ideas on role of manhood in, 100
Eliot, C. W., 70, 71, 197
enunciative modalities, 60, 74, 77
environmental determinism, in *Statistical Atlas* of 1874, 145–146; in Walker's political economy, 177–181
experts, 12, 72–76, 99, 221; role in census enumerations, 132

facts, epistemological status of in argument, 5–6; in modern thought, 27, 29
Farmers' Alliance, 89
feminist scholarship, 9, 84; and theorization of governmentality, 74–76, 129, 222–223
Fitzpatrick, E., 76
Fine, S., 226
fixity, and social reproduction, 14; as exclusionary strategy, 108; contradiction with mobility, 127–130, 223; of workers, 163–171; role in governmentality, 39–40, 222–224; and assortment, 124–130, 149
Fletcher, W. I., 104–105
formation of governmental objects, 43, 59, 60, 77, 221
formation of governmental subjectivity, 43, 60–61, 77, 221
Forum, 53, 65, 66, 68, 80, 193
Foucault, M., and biographical writing, 3–4, 221; and feminism, 20; and geography, 18, 38; and Marxism, 21; and state theory 2, 11, 31–33, 38; on archaeology, 43, 221; on biopower, 19–21; on discipline, 13, 17–19; on gender, 20, 27, 75; on governmentality, 2, 10, 22–25, 188; on power relations, 2; on race, 21, 27; on sexuality, 19–20; scholarship on, 4–5; value of work, 7, 41–42, 188, 221
Fredrickson, G. M., 97
Freud, S., 20

Galaxy, 66
Garfield, J. A., 50–51, 79, 123
gender, differences in treatment of neurasthenia, 88; in governmentality, 221–223; in late nineteenth-century American social order, 84–85, 89; in US census schedules, 57; in Walker's educational thinking, 196–208; in Walker's political economy, 170–171, 174–176,

gender (*cont.*)
  181, 182–183, 185–186; in Walker's view of knowledge-gathering, 129–130; of census office personnel, 139–141
general history, 4, 41, 228
George, H., 163
Germany, significance of for American post-secondary education, 70–71, 201
Giddens, A., 1, 2, 225
Gilded Age, 2–3, 33, 36, 49, 51, 60, 97, 163, 226
Gilman, D. C., 70, 73, 82, 132–133, 138, 139, 182, 196, 197, 201, 203
Godkin, E. L., 53, 62
Government Printing Office, 44–48
governmentality, 10, 22–25, 222–224; and Federalism, 192–194; and liberal political economy, 227–228; and masculinity, 20, 74, 82–83, 85, 222–223; and space, 10, 13, 97, 114–116, 221, 223–224; and state formation, 14, 23, 24, 77, 224–228; as discursive formation, 25–28, 41–42, 60–61, 76–77, 222; geography of discursive formation, 67, 71; inherently political character of, 114–117; place in Foucault's work, 10, 22–25; relation to sovereignty and discipline, 23
Grant, U. S., 81
grids of reference, 117–120
grids of specification, 12, 43, 50, 56–59, 61, 108, 226, in *Statistical Atlas* of 1874, 143–144

Hall, G. S., 94, 201
Hall, P. F., 215, 217
Hamilton, A., 192
Hancock, W. S., 94, 99
Hannah, M., on administration of Oglala Lakota, 7; on space and power, 7–8
Harpers', 53, 65, 66
Harrison, F., 167
Harvard University, 53, 62, 70, 71, 96, 181, 197
Hayes, R. B., 136
Haymarket bombing, 175
Heilbroner, R., 162
Hilkey, J., 86
Hill, J., 216
Hoar, G. F., 79
Hofstadter, R., 90, 91[n22]
Holmes, O. W., 97, 107
Holt [first name unknown] (Walker's first female secretary), 102
homosexuality, 87–88, 91
homosocial organizations, 91–92
Hoyt, J. W., 202

Hull House, 74–75
Hunt's Merchant's Magazine and Statistical Review, 47, 55

immigration, as target of Walker's regulatory program, 208–219; in *Statistical Atlas* of 1874, 153; psychological interpretation of Walker's views on, 109–110; role in Walker's political economy, 176–187
immigrant neighborhoods, difficulty of census enumeration in, 124–129
Immigration Restriction Act, 92, 218
Immigration Restriction League, 215–218
impartiality, importance as character trait, 51, 63–64, 72, 81, 99–100
indexes, as sign of growing volume of knowledge, 67–68
industrial unionism, 92
International Review, 65, 66, 80
invisible hand, 24

Jarvis, E., 55, 62
Jefferson, T., 146, 148, 193, 197
Jessop, B., 33
Jevons, W. S., 162
Johns Hopkins University, 62, 70, 71, 75, 82, 100, 132, 196–197, 203

Keller, M., 35, 36, 97, 226
Kelley, F., 74, 75
Kennedy, J. C. G., 54, 79, 130, 137
Kimmel, M., 86, 87, 96
Knights of Labor, 89

labor movement, 89, 175
labor productivity, 171–176
laissez faire, 27, 35, 36, 74, 81, 163, 165, 166, 167, 191, 220, 226, 227
Lamarckism (or neo-), 177, 179, 180, 208
Latour, B., 4, 13, 28, 39, 113
Le Corbusier, 125
Lefebvre, H., 225
legibility, as goal of systems of control, 115, 119; obstacles to in New York City, 125–126; as criterion for deciding what to observe, 133
liberalism (in political economy), 23–24; and governmentality, 227–228
Lippincott's, 65, 80
literary clubs, importance as forum for cultural elite, 65
Livermore, T., 204
Livingstone, D., 177, 181
Lodge, H. C., 216
Lutz, T., 88

madness, 43, 56
magazines, circulations of major monthlies, 66–67; decline of monthlies, 71–72; importance in disseminating census results, 47; monthlies as fora for elite discourse, 65–69
Malthus, T., 161
manhood, American culture of, 12, 61, 82–83; and governmentality, 222–223; and intellect, 91; and poverty, 90; crisis of, 84–93; destructive effects of ideals, 91; ideals codified in success manuals, 89–90; in Walker's educational philosophy, 100, 196–197, 203–204; in Walker's political economy, 160, 174–176, 181, 182–183, 185–187; in Walker's regulatory program, national, 86, 182, 183, 186; passionate, 90–91; self-made, 87–89; value of anger in, 94; Walker's military model of, 13, 96–104, 140–141
Mann, H., 200
Mann, M., 1, 2, 25, 38–39, 68, 225
marginal economics, 160–161
Marshall, A., 94, 162
Marshall, J., 193
Marx, K., 214; scholarship on, 5; as political economist, 163, 164
Marxism, 9, 224
masculinity, *see* manhood
Massachusetts Institute of Technology, 80, 82, 100, 101, 102, 134, 197, 201, 202, 203, 204, 205, 206, 207, 208, 214
McLaren, A., 88
McNulty, P., 173
military virtues, 94
Mill, J. S., 162, 163
Miller, P., 22, 26, 221
mobility, contradiction with fixity, 127–130, 223; of workers, 14, 163–171, 213; role in governmentality, 39–40, 213–215, 222–224; and abstraction, 117–123, 127–130, 149
Moscow Imperial Technical School, 202
Mott, F. L., 64
Mrozek, D., 98
Munroe, J. P., 93, 94, 95, 96, 100, 101, 102
Murdoch, J., 8

Nation, 53, 65, 66, 72, 80
nationalism, Walker on origins of American, 193
nativism, 92; in Walker's political economy, 174–187
Nelson, D., 86
neurasthenia, 87–89
newspapers, role in distribution of census results, 48–49

Newton, B., 162, 173
New York City, as center of magazine publishing, 65, 67; as difficult city to enumerate, 124–128
New York Times, 48
Niles Weekly Register, 47
normalizing judgment, 14, 40, 60, 148, 186, 188
North, S. N. D., 217
North American Review, 53, 65, 66, 68, 72, 80, 210
Nyerere, J., 125

objectivity, 5–6, 14, 59, 60, 135
observation (governmental), 13, 39, 60, 113, 117, 148, 186, 188
Osborne, T., 76

panopticon, 13, 18, 136
Parsons, E. C., 74
patronage, as obstacle to centralized control of census-taking, 133–134; role of in nineteenth-century American state, 35, 37–39, 52–56
pauperism, as touchstone for social economy, 26; Walker on, 182
Philadelphia, as easy city to enumerate, 124–125
Philadelphia Centennial Exhibition, 81, 95, 202
physiocratic theory, 23
Pierce, B. F., 69
political economy, history of, 160–164; role in governmentality, 21, 23–24; Walker's, 14–15, 164–186
Poole, W. F., 65, 67, 104–106
Poole's Index of Periodical Literature, 65, 66, 104
Poovey, M., 6, 27, 28–29, 37
Popular Science Monthly, 66
populism, 92
population growth, 57–58, 182–187, 194
Porter, T., 11, 29–30, 73
Poulantzas, N., 33
Powell, J. W., 29, 226
Prévost, J.-G., 216
Princeton Review, 65, 80
Procacci, G., 26–27, 76
professional associations, 69–71
Progressive Era, 1, 12, 36, 73, 83, 98, 99, 221, 227
protectionism, Walker on, 194–195
Putnam's, 65

Rabinow, P., 225–226
race and ethnicity, 9; as grid of specification, 57; critical studies of, 9; in late nineteenth-century American social order, 85; in

race and ethnicity (*cont.*)
  Statistical Atlas of 1874, 144, 146; in
    Walker's political economy, 174–187; in
    Walker's regulatory program, 208–219; role
    in colonial administration, 114; role in
    exclusionary psychology, 108–109;
    statistics of, 51, 216–218
regulation, as moment in cycle of social
    control, 40, 186, 188–189; characteristics of
    governmental, 189, 190, 194, 224; Walker's
    views on, 15, 189–219
reformers (Republican), 53–54, 69–70
residual claimant theory of wage
    determination, 172–176
Review of Reviews, 67, 103
Ricardo, D., 161, 162, 163
Richards, E., 207
Riis, J., 125–128
Robertson, R., 161
Rodgers, W. B., 102
Roll, E., 96, 161, 162
Roosevelt, T., 74, 88, 98, 217
Rose, G., 129
Rose, N., 22, 26, 76, 221
Ross, D., 70, 73, 96, 148
Rossiter, C., 216, 217
Rotundo, E. A., 84–85, 86, 89, 90, 96
Runkle, J. D., 102, 202

savoir, 26–27, 43
Schumpeter, J., 162
Schurz, C., 53, 62
Scott, J. C., 4, 13, 115, 119, 125, 130, 203,
    225
Scott, J. W., 84
Scribner's, 65, 66, 80
Seaton, C. W., 50, 102
Senior, N., 162
Shaler, N. S., 62, 100, 181
Shattuck, L., 55
Shils, E., 72
Sibley, D., 108–110
Sidwell, R. J., 78, 81
Silverberg, H., 76
Sklar, K. K., 76
Skowronek, S., 35, 37, 226
Small, A., 75
Smith, A., 23–24, 160–161, 182
Snead, W. T., 67
social body, 24–25, 28, 58, 142, 188, 190, 208,
    219, 221, 223, 224
social constructionism, 10; and biographical
    genre, 3–4, 76–78, 82–83, 110–111; in state
    theory, 224–225
social Darwinism, 90, 166, 195, 216
social economy, 26–27, 75
socialism, Walker on, 190–192

social science, and the organization of
    American society, 97–98; and social work,
    74, origins of, 69–76; professional journals,
    70–71
Solomon, B., 175
space, and power, 7, 17–18, 38–40, 107–110,
    114–117, 208–219, 222–224
spatial politics of census taking, 13, 39–40,
    107–108, 114–141
special commissions, significance of, 68–69
stakeholders, 86, 89, 90, 92, 93
Stanford, L., 197
Stanford University, 197
state formation, 1, 8; contributions of
    governmentality to theory of, 224–228;
    early modern American, 34–36, 68, 77; in
    colonial vs. metropolitan contexts,
    114–117; theoretical issues of 8–9, 31–34,
    36–38, 224–228
Statistical Atlas of 1874, 141–149, 151–158;
    and compilation, 142; basic structure of,
    143–144, 145–146; environmental
    determinism in, 145–146; map on fertility,
    183; purported objectivity of, 142;
    selectivity of, 144–145; social construction
    of "foreigners" in, 143–144, 154; visual
    rhetoric of, 146–147
statistics, advantages for programs of
    governmentality, 28–29; American
    fascination with, 29–30; course at MIT,
    203; role in American public life, 29–31, 63,
    73, 76; role in social science, 73; early
    difficulty receiving education in, 62–63
Stearns, P., 94
Stoughton, E. (Mrs. Walker), 95, 103,
    182–183
structuralism, 41
Sumner, W. G., 195
surfaces of emergence, 12, 43; of American
    governmental discourse, 43–49, 61
symposia, significance of, 68

Takaki, R., 210
Taylor, F., 202
Taylor, Z., 137
telegraph, impact on news distribution, 48
tenements, as obstacles to census taking,
    124–130
territorial mastery, 10, 39–40, 107–110; as
    epistemological project, 114–149; as
    regulatory program, 208–217, 219, 222–224
teutonism, 177, 179
theory, and selectivity, 5–6; role in narrative,
    220
Townshend, K., 95–96
transportation and communication
    infrastructure, 48, 97, 120–121

Turner, F. J., 144, 146, 197

United States Congress, and patronage, 50;
  role in distributing census results, 44–47;
  role in shaping censuses, 50–52
United States Immigration Commission,
  217–218
United States Sanitary Commission, 53
universities, geographical distribution of, 71;
  significance of for American
  governmentality, 71–72

wage fund theory, 164–166
wages, Walker on determination of, 164–176
Walker, A., 78
Walton, G., 161
War of 1812, 193
Ward, N., 8
Ward, R. de C., 215

Warren, C., 215
Washington, G., 192
Washington University, 202
Watson, E., 183, 184, 194
Weaver, W., 51–52
Wells, D. A., 55, 56, 97
White, A., 71, 201
White, L., 53–54
Wiebe, R., 1, 84
Wishart, D., 5–6
women, at MIT, 205–207; in Census Office,
  139–140; in early American social science,
  74–76; in public life, 92; in Walker's
  political economy, 170–171
Woodward, C. M., 202
Wright, C., 36, 55, 56, 73, 97, 134

Yale Review, 215
Yale University, 53, 74, 195, 197

Cambridge Studies in Historical Geography

1 Period and place: research methods in historical geography. *Edited by* ALAN R. H. BAKER and M. BILLINGE

2 The historical geography of Scotland since 1707: geographical aspects of modernisation. DAVID TURNOCK

3 Historical understanding in geography: an idealist approach. LEONARD GUELKE

4 English industrial cities of the nineteenth century: a social geography. R. J. DENNIS*

5 Explorations in historical geography: interpretative essays. *Edited by* ALAN R. H. BAKER and DEREK GREGORY

6 The tithe surveys of England and Wales. R. J. P. KAIN and H. C. PRINCE

7 Human territoriality: its theory and history. ROBERT DAVID SACK

8 The West Indies: patterns of development, culture and environmental change since 1492. DAVID WATTS*

9 The iconography of landscape: essays in the symbolic representation, design and use of past environments. *Edited by* DENIS COSGROVE and STEPHEN DANIELS*

10 Urban historical geography: recent progress in Britain and Germany. *Edited by* DIETRICH DENECKE *and* GARETH SHAW

11 An historical geography of modern Australia: the restive fringe. J. M. POWELL*

12 The sugar-cane industry: an historical geography from its origins to 1914. J. H. GALLOWAY

13 Poverty, ethnicity and the American city, 1840–1925: changing conceptions of the slum and ghetto. DAVID WARD*

14 Peasants, politicians and producers: the organisation of agriculture in France since 1918. M. C. CLEARY

15 The underdraining of farmland in England during the nineteenth century. A. D. M. PHILLIPS

16 Migration in Colonial Spanish America. *Edited by* DAVID ROBINSON

17 Urbanising Britain: essays on class and community in the nineteenth century. *Edited by* GERRY KEARNS and CHARLES W. J. WITHERS

18 Ideology and landscape in historical perspective: essays on the meanings of some places in the past. *Edited by* ALAN R. H. BAKER and GIDEON BIGER

19 Power and pauperism: the workhouse system, 1834–1884. FELIX DRIVER

20 Trade and urban development in Poland: an economic geography of Cracow from its origins to 1795. F. W. CARTER

21 An historical geography of France. XAVIER DE PLANHOL

22 Peasantry to capitalism: Western Östergötland in the nineteenth century.
GÖRAN HOPPE and JOHN LANGTON

23 Agricultural revolution in England: the transformation of the agrarian economy,
1500–1850. MARK OVERTON*

24 Marc Bloch, sociology and geography: encountering changing disciplines.
SUSAN W. FRIEDMAN

25 Land and society in Edwardian Britain. BRIAN SHORT

26 Deciphering global epidemics: analytical approaches to the disease records of
world cities, 1888–1912. ANDREW CLIFF, PETER HAGGETT and MATTHEW
SMALLMAN-RAYNOR*

27 Society in time and space: a geographical perspective on change.
ROBERT A. DODGSHON*

28 Fraternity among the French peasantry: sociability and voluntary associations in
the Loire valley, 1815–1914. ALAN R. H. BAKER

29 Imperial Visions: nationalist imagination and geographical expansion in the
Russian Far East, 1840–1865. MARK BASSIN

30 Hong Kong as a global metropolis. DAVID M. MEYER

31 English seigniorial agriculture, 1250–1450. BRUCE M. S. CAMPBELL

*Titles marked with an asterisk * are available in paperback.*

Printed in the United States
By Bookmasters